Lecture Notes in Artificial Intelligence 7081

Subseries of Lecture Notes in Computer Science

Friedhelm Schwenker
Edmondo Trentin (Eds.)

Partially Supervised Learning

First IAPR TC3 Workshop, PSL 2011
Ulm, Germany, September 15-16, 2011
Revised Selected Papers

 Springer

Series Editors

Randy Goebel, University of Alberta, Edmonton, Canada
Jörg Siekmann, University of Saarland, Saarbrücken, Germany
Wolfgang Wahlster, DFKI and University of Saarland, Saarbrücken, Germany

Volume Editors

Friedhelm Schwenker
University of Ulm
Institute of Neural Information Processing
89069 Ulm, Germany
E-mail: friedhelm.schwenker@uni-ulm.de

Edmondo Trentin
University of Siena
DII – Dipartimento di Ingegneria dell'Informazione
Via Roma 56, 53100 Siena, Italy
E-mail: trentin@dii.unisi.it

ISSN 0302-9743   e-ISSN 1611-3349
ISBN 978-3-642-28257-7   ISBN 978-3-642-28258-4 (eBook)
DOI 10.1007/978-3-642-28258-4
Springer Heidelberg Dordrecht London New York

Library of Congress Control Number: 2012930867

CR Subject Classification (1998): I.2.6, I.2, I.5, I.4, H.3, F.2.2, J.3

LNCS Sublibrary: SL 7 – Artificial Intelligence

Typesetting: Camera-ready by author, data conversion by Scientific Publishing Services, Chennai, India

Printed on acid-free paper

Springer is part of Springer Science+Business Media (www.springer.com)

Preface

Partially supervised learning (PSL) is a rapidly evolving area of machine learning. In many applications unlabeled data may be relatively easy to collect, whereas labeling these data is difficult, expensive or/and time consuming as it needs the effort of human experts. PSL is a general framework for learning with labeled and unlabeled data, for instance, in classification, it is assumed that each learning sample consists of a feature vector and some information about its class. In the PSL framework this information might be a crisp label, or a label plus a confidence value, or it might be an imprecise and/or uncertain soft label defined through a certain type of uncertainty model (fuzzy, Dempster–Shafer), or it might be that information about a class label is not available.

The PSL framework thus generalizes many kinds of learning paradigms including supervised and unsupervised learning, semi-supervised learning for classification and regression, transductive learning, semi-supervised clustering, policy learning in partially observable environments, and many others. Therefore PSL methods and algorithms are of great interest in both practical applications and theory. Research in the field of PSL is still in its early stages and has great potential for further growth, thus leaving plenty of room for further development.

This First IAPR-TC3 Workshop on Partially Supervised Learning (PSL 2011), whose proceedings are presented in this volume, endeavored to bring together recent novel research in this area and to provide a forum for further discussion. The workshop was held at the University of Ulm (http://neuro.informatik.uni-ulm.de/PSL2011/), Germany, during September 15–16, 2011. It was supported by the International Association of Pattern Recognition (IAPR) and by the IAPR Technical Committee on Neural Networks and Computational Intelligence (TC 3). IAPR-TC3 is one of the 20 Technical Committees of IAPR, focusing on the application of computational intelligence to pattern recognition.

PSL 2011 focused on a number of different topics, covering methodological issues as well as real-world applications of PSL. The main methodological issues were: combination of supervised and unsupervised learning; diffusion learning; semi-supervised classification, regression, and clustering; learning with deep architectures; active leaning; PSL with vague, fuzzy, or uncertain teaching signals; PSL in multiple classifier systems and ensembles; PSL in neural nets, machine learning, or statistical pattern recognition; PSL in cognitive systems. Applications of PSL included: image and signal processing; multi-modal information processing; sensor/information fusion; human–computer interaction; data mining and Web mining; forensic anthropology; bioinformatics.

The workshop featured regular oral presentations and a poster session, plus three brilliant invited speeches, namely: "Unlabeled Data and Multiple Views," delivered by Zhi-Hua Zhou (LAMDA Group, National Key Laboratory for

Novel Software Technology, Nanjing University, Nanjing, China); "Online Semi-Supervised Ensemble Updates for fMRI Data," delivered by Catrin Plumpton (University of Wales, Bangor, UK); and, "How Partially Supervised Learning Can Facilitate and Enhance User State Analysis in Naturalistic HCI," delivered by Stefan Scherer (Trinity College Dublin, Ireland). It is our firm conviction that all the papers in this book are of high quality and significance to the area of PSL. We sincerely hope that readers of this volume may, in turn, enjoy it and get inspired from the different contributions.

We would like to acknowledge the fact that the organization of the workshop made its first steps within the framework of the Vigoni Project for international exchanges between the universities of Siena (Italy) and Ulm (Germany). Also, we wish to acknowledge the generosity of the PSL 2011 sponsors: IAPR, IAPR-TC3, the University of Ulm which hosted this event, and the Transregional Collaborative Research Centre SFB/TRR 62 *Companion-Technology for Cognitive Technical Systems* for generous financial support. We are grateful to all the authors who submitted a paper to the workshop, since their efforts and invaluable contributions led to a great event. Special thanks to the local organization crew based in Ulm, namely, Michael Glodek, Martin Schels, Miriam K. Schmidt and Sascha Meudt. The contribution from the members of the Program Committee in promoting the event and reviewing the papers is gratefully acknowledged. Finally, we wish to express our gratitude to Springer for publishing these proceedings within their LNCS/LNAI series, and for their constant support.

October 2011

Friedhelm Schwenker
Edmondo Trentin

Organization

Organizing Committee

Friedhelm Schwenker   University of Ulm, Germany
Edmondo Trentin   University of Siena, Italy

Program Committee

Erwin Bakker (The Netherlands)
Yoshua Bengio (Canada)
Hendrik Blockeel
    (Belgium/The Netherlands)
Paolo Frasconi (Italy)
Neamat El Gayar (Egypt)
Marco Gori (Italy)
Markus Hagenbuchner (Australia)
Barbara Hammer (Germany)
Tom Heskes (The Netherlands)
Lakhmi Jain (Australia)
Manfred Jaeger (Denmark)
Cheng-Lin Liu (China)

Nik Kasabov (New Zealand)
Marco Maggini (Italy)
Simone Marinai (Italy)
Günther Palm (Germany)
Lionel Prevost (France)
Alessandro Sperduti (Italy)
Fabio Roli (Italy)
Ah Chung Tsoi (Hong Kong)
Michel Verleysen (Belgium)
Terry Windeatt (UK)
Ian Witten (New Zealand)

Local Arrangements

Miriam K. Schmidt
Michael Glodek
Sascha Meudt
Martin Schels

Sponsoring Institutions

University of Ulm, Germany
University of Siena, Italy
Transregional Collaborative Research Centre SFB/TRR 62 *Companion-Technology for Cognitive Technical Systems*
International Association for Pattern Recognition (IAPR)

Table of Contents

Unlabeled Data and Multiple Views

Zhi-Hua Zhou

National Key Laboratory for Novel Software Technology
Nanjing University, Nanjing 210093, China
zhouzh@lamda.nju.edu.cn

Abstract. In many real-world applications there are usually abundant unlabeled data but the amount of labeled training examples are often limited, since labeling the data requires extensive human effort and expertise. Thus, exploiting unlabeled data to help improve the learning performance has attracted significant attention. Major techniques for this purpose include semi-supervised learning and active learning. These techniques were initially developed for data with a single *view*, that is, a single *feature set*; while recent studies showed that for multi-view data, semi-supervised learning and active learning can amazingly well. This article briefly reviews some recent advances of this thread of research.

1 Introduction

Traditional supervised learning approaches try to learn from *labeled* training examples, i.e., training examples with ground-truth labels given in advance. In many real-world tasks, however, there are often abundant *unlabeled* data but limited amount of labeled training examples. Simply neglecting the unlabeled data would waste useful information, while learning only from the limited labeled data would be difficult to achieve strong generalization performance. Thus, it is natural that exploiting unlabeled data to help improve learning performance, especially when there are just a few training examples, has attracted significant attention during the past decade.

Major techniques for this purpose include *semi-supervised learning* and *active learning*. Semi-supervised learning [5, 30, 28] tries to exploit unlabeled data in addition to labeled data automatically, without human intervention; while active learning [17] assumes interaction with an *oracle*, usually human experts, by trying to minimizing the number of queries on ground-truth labels for constructing a strong learning model. Semi-supervised learning can be divided further into *pure* semi-supervised learning which takes an open-world assumption that the trained model may be applied to unseen unlabeled data, and *transductive learning* which adopts a closed-world assumption that the test instances are exactly the given unlabeled data. The idea of transductive learning can be traced back to [20], where it was argued that we do not need to optimize the learning performance on the whole instance space if we only care the generalization performance on a specific set of test instances.

F. Schwenker and E. Trentin (Eds.): PSL 2011, LNAI 7081, pp. 1–7, 2012.
© Springer-Verlag Berlin Heidelberg 2012

Data in many tasks have only a single *view*, i.e., a single feature set, and each instance is described by a single feature vector in such situations. However, there are also many real-world tasks where the data have multiple views, i.e., multiple feature sets, and each instance is described by multiple feature vectors in different feature spaces simultaneously. For example, a web page can be classified based on information appearing in the web page itself, or based on anchor texts pointing to this web page; thus, features describing the information in the web page itself constitute the first view, while features describing the information in the anchor texts constitute the second view. Another example is multimedia data, where text features, image features and audio features constitute three different views, respectively. Formally, a single-view example appears as $(\boldsymbol{x}_i, y_i)$ where $\boldsymbol{x}_i$ is the instance and y_i is the class label; while a multi-view example appears as $([\boldsymbol{x}_{i1}, \boldsymbol{x}_{i2}], y_i)$ where $[\boldsymbol{x}_{i1}, \boldsymbol{x}_{i2}]$ is an instance pair in different views (e.g., $\boldsymbol{x}_{i1}$ is a text feature vector while $\boldsymbol{x}_{i2}$ is an image feature vector). Rather than simply concatenating $\boldsymbol{x}_{i1}$ and $\boldsymbol{x}_{i2}$ into a single instance, *multi-view learning* deals with multi-view data by exploiting the views.

Semi-supervised learning and active learning techniques were initially developed for single-view data. It has been found that, however, for multi-view data, semi-supervised learning and active learning can work amazingly well. This article briefly reviews some recent advances of this thread of research.

2   Semi-supervised Learning and Multi-view

Among mainstream semi-supervised learning techniques, the disagreement-based approaches are particularly interesting. These approaches train multiple learners for the task and exploit the disagreements among the learners during the semi-supervised process [28]. A representative is the co-training approach [3] which works with two views. This approach trains a classifier from each view, respectively, using the original labeled data. Then, each classifier selects and labels some highly-confident unlabeled instances to refine its peer classifier. The whole process repeats until no classifier changes or a pre-set number of learning rounds have been executed.

Such a learning process is simple yet effective, and it has many variants and applications [28]. Theoretically, Blum and Mitchell [3] proved that if the two views are "sufficient and redundant" (i.e., each view contains sufficient information for constructing a strong classifier while the two views are conditionally independent given the class label), the predictive accuracy of an initial weak classifier can be boosted to arbitrarily high using unlabeled data by co-training. Dasgupta et al. [7] showed that the generalization error of co-training is upper-bounded by the disagreement between the two classifiers. In real-world tasks, however, the requirement of sufficient and redundant views is too luxury. Actually, even for the motivating example of web page classification task given in [3], it is arguable that whether the requirement holds or not. Thus, researchers tried to find relaxed conditions for co-training to work.

Abney [1] showed that the two views are not needed to be conditionally independent, and a "weak independence" assumption is sufficient for co-training to work. Balcan et al. [2] proved that even the weak independence is not needed if PAC learners can be obtained on each view, and a weaker assumption of "expansion" of the underlying data distribution is sufficient for co-training to work. All the above analyses assumed two views. Wang and Zhou [21] disclosed that for PAC learners, the key for co-training-style approaches is the existence of a "large difference" between the two learners, while it is unimportant whether the difference is achieved by using two views or from other channels. This result provides theoretical support to single-view variants of co-training which work well without two views by training the two learners using different learning algorithms [9], different parameter configurations [27, 10], etc.

As introduced above, more and more relaxed *sufficient conditions* for co-training have been discovered; however, the *sufficient and necessary condition* remained unknown for over ten years. Recently, through establishing a connection between the two mainstream semi-supervised learning approaches, that is, disagreement-based and graph-based approaches, Wang and Zhou [24] addressed this problem. They showed that the co-training process is equivalent to a combinative label propagation process over graphs corresponding to the two views, and thus, sufficient and necessary conditions for co-training were discovered by analyzing the properties of the corresponding graphs under different situations. Wang and Zhou [24] also proved a *necessary condition*, which discloses that the existence of two views is not really needed for co-training-style approaches.

Now it is known that multi-view is neither necessary [24] nor "tightly" sufficient [21] for co-training-style approaches; however, when the data have multiple views, amazing performances can be achieved. For example, Zhou et al. [29] showed that, with sufficient and redundant views, it is possible to execute an effective semi-supervised learning with a single labeled training example, owing to helpful information contained in the correlation between the two views.

3   Active Learning and Multi-view

Active learning generally tries to query the labels of unlabeled *informative* instances (e.g., [18]) or *representative* instances (e.g., [6]). Recently there are some proposals of querying on *informative and representative* unlabeled instances (e.g., [11]). Those principles can be accomplished in different ways, leading to different active learning approaches. A simple yet effective multi-view active learning approach, co-testing [14], trains two classifiers each from one view and then picks their most disagreed unlabeled instance to query, with the intuition that the most disagreed unlabeled instance would be the most informative for improving learning performance.

Theoretically there are two situations of active learning; that is, *realizable* active learning where the data can be perfectly separated by a hypothesis in the hypothesis class, and *non-realizable* active learning where the data cannot be perfectly separated by any hypothesis in the hypothesis class because of noise. For the realizable case, many studies showed that *exponential* improvement in sample complexity can be achieved by active learning. Wang and Zhou [22] proved that an multi-view active learning approach can also improve the sample complexity remarkably in realizable case.

The realizability assumption, however, rarely holds in real practice, and the non-realizable case is more important since it is closer to real setting. Kääriäinen [12] showed that the lower bound of general non-realizable active learning is in the same order as the upper bound of *passive learning* (i.e., common supervised learning), implying that active learning in the non-realizable case is not as helpful as that in the realizable case if nothing is known about the noise model. In analyses on non-realizable active learning, the Tsybakov noise model [19] becomes more and more popular. It is known that exponential improvement in sample complexity is achievable with *bounded* Tsybakov noise, but for *unbounded* Tsybakov noise which is more closer to real settings, several researchers such as Castro and Nowak [4] concluded that it is hard to achieve exponential improvement, or in other words, active learning would not be remarkably helpful. Recently, Wang and Zhou [23] proved that an multi-view active learning approach can exponentially improve the sample complexity in non-realizable case with unbounded Tsybakov noise. This is a good news, implying that active learning is possible to help remarkably if specific data properties are adequately considered and exploited.

It is not difficult to combine multi-view active learning with semi-supervised learning. For example, Muslea et al. [15] combined co-testing with co-EM [16], a probabilistic variant of co-training, where co-EM iteratively learned two classifiers each from one view by exploiting unlabeled data, and the unlabeled instances disagreed by the two classifiers were selected to query by co-testing. Empirical studies showed that such an approach performed better than semi-supervised learning. Zhou et al. [25] proposed a single-view active semi-supervised learning approach for content-based image retrieval. They generated two learners from labeled images using different parameter configurations. Each learner attempts to assign a rank to unlabeled images in the imagebase, and then passes some irrelevant images with high confidence to its peer as additional negative examples. The two learners are updated and such a process repeats. At the meanwhile, rather than passively waiting for user feedback, a pool of images is actively prepared for the user to give feedback. The pool is composed of images on which the two learners are both with high confidences but disagree, or both with low confidences. The whole process leads to the *active semi-supervised relevance feedback* scheme which is useful for information retrieval tasks. Theoretically, Wang and Zhou [22] proved that an multi-view active semi-supervised learning approach is able to exponentially improve sample complexity in contrast to pure semi-supervised learning.

4   About the Views

Different assumptions can be made for the views, from the possibly weakest that "each view contains information for training weak classifiers that are slightly better than random guess", to the possibly strongest that the views are "sufficient and redundant".

View split, i.e., splitting a single view into multiple views, is a possible solution for applying multi-view approaches to single-view data. It was shown in [16] that for data with a lot of redundant features, such as text data, a random split of the features is able to generate two views to enable standard co-training. It is evident, however, that a random split would not work in most cases. Du et al. [8] tried several heuristics for view split and found that all heuristics failed with insufficient labeled data. The necessary condition of co-training given in [24] suggested that among all potential view splits, the one which enables the most unlabeled instances connect with labeled examples in the combinative graph is preferred; this was empirically verified in [24] and might give inspiration to develop sound practical view split approaches.

Most previous studies on multi-view learning focused on two views, possibly owing to the fact that less data sets with more than two views are publicly available. With the increasing demand of multimedia data analysis, data with more than two views become more accessible. Extending two-view approaches to more views, however, is not trivial. This is because helpful information are concealed in the relations between the views, while the relations become more complicated with more views. A simple "view-invariant" approach is to train one learner from each view, and then let the learners exploit unlabeled data through the strategy of *majority teach minority*. This strategy has been found effective in single-view multi-learner semi-supervised learning approaches tri-training [26] and co-forest [13], and is expected to be helpful on multi-view data. Furthermore, such a semi-supervised learning process would be easy to combine with committee-based active learning approaches.

5   Conclusion

This article briefly reviews some recent advances in exploiting unlabeled data with multiple views. Now it is known that multi-view is not really needed for disagreement-based semi-supervised learning approaches such as co-training; however, given adequate multiple views, amazing performances such as semi-supervised learning with a single labeled example becomes possible. Multi-view also enables exponential improvement of sample complexity for non-realizable active learning with unbounded Tsybakov noise. Overall, multi-view brings great potential of interesting new findings and strong learning approaches for exploiting unlabeled data.

Acknowledgments. This article summarizes the keynote given at the IAPR Workshop on Partially Supervised Learning (PSL), Ulm, Germany, in September 2011. The author was supported by the National Fundamental Research Program (2010CB327903) and the National Science Foundation of China (61073097, 61021062).

References

1. Abney, S.: Bootstrapping. In: Proceedings of the 40th Annual Meeting of the Association for Computational Linguistics, Philadelphia, PA, pp. 360–367 (2002)
2. Balcan, M.-F., Blum, A., Yang, K.: Co-training and expansion: Towards bridging theory and practice. In: Saul, L.K., Weiss, Y., Bottou, L. (eds.) Advances in Neural Information Processing Systems, vol. 17, pp. 89–96. MIT Press, Cambridge, MA (2005)
3. Blum, A., Mitchell, T.: Combining labeled and unlabeled data with co-training. In: Proceedings of the 11th Annual Conference on Computational Learning Theory, Madison, WI, pp. 92–100 (1998)
4. Castro, R.M., Nowak, R.D.: Minimax bounds for active learning. IEEE Transactions on Information Theory 54(5), 2339–2353 (2008)
5. Chapelle, O., Schölkopf, B., Zien, A. (eds.): Semi-Supervised Learning. MIT Press, Cambridge, MA (2006)
6. Dasgupta, S., Hsu, D.: Hierarchical sampling for active learning. In: Proceedings of the 25th International Conference on Machine Learning, Helsinki, Finland, pp. 208–215 (2008)
7. Dasgupta, S., Littman, M., McAllester, D.: PAC generalization bounds for co-training. In: Dietterich, T.G., Becker, S., Ghahramani, Z. (eds.) Advances in Neural Information Processing Systems, vol. 14, pp. 375–382. MIT Press, Cambridge, MA (2002)
8. Du, J., Ling, C.X., Zhou, Z.-H.: When does co-training work in real data? IEEE Transactions on Knowledge and Data Engineering 23(5), 788–799 (2010)
9. Goldman, S., Zhou, Y.: Enhancing supervised learning with unlabeled data. In: Proceedings of the 17th International Conference on Machine Learning, San Francisco, CA, pp. 327–334 (2000)
10. Guo, Q., Chen, T., Chen, Y., Zhou, Z.-H., Hu, W., Xu, Z.: Effective and efficient microprocessor design space exploration using unlabeled design configurations. In: Proceedings of the 22nd International Joint Conference on Artificial Intelligence, Barcelona, Spain, pp. 1671–1677 (2011)
11. Huang, S.-J., Jin, R., Zhou, Z.-H.: Active learning by querying informative and representative examples. In: Lafferty, J., Williams, C.K.I., Shawe-Taylor, J., Zemel, R.S., Culotta, A. (eds.) Advances in Neural Information Processing Systems, vol. 23, pp. 892–900. MIT Press, Cambridge, MA (2010)
12. Kääriäinen, M.: Active learning in the non-realizable case. In: Proceedings of the 44th Annual Meeting of the Association for Computational Linguistics, Sydney, Australia, pp. 63–77 (2006)
13. Li, M., Zhou, Z.-H.: Improve computer-aided diagnosis with machine learning techniques using undiagnosed samples. IEEE Transactions on Systems, Man and Cybernetics - Part A: Systems and Humans 37(6), 1088–1098 (2007)
14. Muslea, I., Minton, S., Knoblock, C.A.: Selective sampling with redundant views. In: Proceedings of the 17th National Conference on Artificial Intelligence, Austin, TX, pp. 621–626 (2000)

15. Muslea, I., Minton, S., Knoblock, C.A.: Active + semi-supervised learning = robust multi-view learning. In: Proceedings of the 19th International Conference on Machine Learning, Sydney, Australia, pp. 435–442 (2002)
16. Nigam, K., Ghani, R.: Analyzing the effectiveness and applicability of co-training. In: Proceedings of the 9th ACM International Conference on Information and Knowledge Management, Washington, DC, pp. 86–93 (2000)
17. Settles, B.: Active learning literature survey. Technical Report 1648, Department of Computer Sciences, University of Wisconsin at Madison, Wisconsin, WI (2009), http://pages.cs.wisc.edu/~bsettles/pub/settles.activelearning.pdf
18. Tong, S., Chang, E.: Support vector machine active learning for image retrieval. In: Proceedings of the 9th ACM International Conference on Multimedia, Ottawa, Canada, pp. 107–118 (2001)
19. Tsybakov, A.: Optimal aggregation of classifiers in statistical learning. Annals of Statistics 32(1), 135–166 (2004)
20. Vapnik, V.N.: Statistical Learning Theory. Wiley, New York (1998)
21. Wang, W., Zhou, Z.-H.: Analyzing Co-training Style Algorithms. In: Kok, J.N., Koronacki, J., Lopez de Mantaras, R., Matwin, S., Mladenič, D., Skowron, A. (eds.) ECML 2007. LNCS (LNAI), vol. 4701, pp. 454–465. Springer, Heidelberg (2007)
22. Wang, W., Zhou, Z.-H.: On multi-view active learning and the combination with semi-supervised learning. In: Proceedings of the 25th International Conference on Machine Learning, Helsinki, Finland, pp. 1152–1159 (2008)
23. Wang, W., Zhou, Z.-H.: Multi-view active learning in the non-realizable case. In: Lafferty, J., Williams, C.K.I., Shawe-Taylor, J., Zemel, R.S., Culotta, A. (eds.) Advances in Neural Information Processing Systems, vol. 23, pp. 2388–2396. MIT Press, Cambridge, MA (2010)
24. Wang, W., Zhou, Z.-H.: A new analysis of co-training. In: Proceedings of the 27th International Conference on Machine Learning, Haifa, Israel, pp. 1135–1142 (2010)
25. Zhou, Z.-H., Chen, K.-J., Dai, H.-B.: Enhancing relevance feedback in image retrieval using unlabeled data. ACM Transactions on Information Systems 24(2), 219–244 (2006)
26. Zhou, Z.-H., Li, M.: Tri-training: Exploiting unlabeled data using three classifiers. IEEE Transactions on Knowledge and Data Engineering 17(11), 1529–1541 (2005)
27. Zhou, Z.-H., Li, M.: Semi-supervised regression with co-training style algorithms. IEEE Transactions on Knowledge and Data Engineering 19(11), 1479–1493 (2007)
28. Zhou, Z.-H., Li, M.: Semi-supervised learning by disagreement. Knowledge and Information Systems 24(3), 415–439 (2010)
29. Zhou, Z.-H., Zhan, D.-C., Yang, Q.: Semi-supervised learning with very few labeled training examples. In: Proceedings of the 22nd AAAI Conference on Artificial Intelligence, Vancouver, Canada, pp. 675–680 (2007)
30. Zhu, X.: Semi-supervised learning literature survey. Technical Report 1530, Department of Computer Sciences, University of Wisconsin at Madison, Madison, WI (2006), http://www.cs.wisc.edu/~jerryzhu/pub/ssl_survey.pdf

Online Semi-supervised Ensemble Updates for fMRI Data

Catrin O. Plumpton

School of Computer Science
Bangor University
Dean Street, LL57 1UT, United Kingdom
c.o.plumpton@bangor.ac.uk

Abstract. Advances in Eelectroencephalography (EEG) and functional magnetic resonance imaging (fMRI) have opened up the possibility for real time data classification. A small amount of labelled training data is usually available, followed by a large stream of unlabelled data. Noise and possible concept drift pose a further challenge. A fixed pre-trained classifier may not always work. One solution is to update the classifier in real-time. Since true labels are not available, the classifier is updated using the predicted label, a method called naive labelling. We propose to use classifier ensembles in order to counteract the adverse effect of 'run-away' classifiers, associated with naive labelling. A new ensemble method for naive labelling is proposed. The label taken to update each member-classifier is the ensemble prediction. We use an fMRI dataset to demonstrate the advantage of the proposed method over the fixed classifier and the single classifier updated through naive labelling.

Keywords: Semi-supervised learning, random subspace ensemble, fMRI.

1   Introduction

Functional Magnetic Resonance Imaging (fMRI) data provides a spatially accurate account of activity within the human brain. Analysis of this activity allows valuable insight into the architecture of the mind, and into human behaviour.

By using multivariate techniques such as classification, we can train a classifier to predict which stimuli are being presented to the subject, based upon the acquired fMRI volume images. Traditionally, fMRI analyses are carried out offline, once scanning has been completed. The question most often asked is which brain regions are responsible for different behaviours and emotions? Responses of different brain regions can also be linked to illness such as depression or schizophrenia. fMRI analysis pre and post treatment can be then used as a measure of success of treatments [15], [35], [16].

By applying online and real time classification to the data as it is collected, neuroscientists acquire feedback during the course of the trial. This facilitates

F. Schwenker and E. Trentin (Eds.): PSL 2011, LNAI 7081, pp. 8–18, 2012.

neurofeedback and self-regulation type experiments, or for the controller to adjust the stimuli in accordance with the subject's reaction. Typical real-time experiments include using the mind to control an object, such as a pendulum [6], or a ball in a maze [37], [28]. Similar techniques have been used to allow subjects to form words using a character map [5]. Other real time experiments include self-control of specific brain regions, for example those involved in pain perception [4] or sadness [33]. Such experiments are typically undertaken in a closed loop brain computer interface (BCI) [8], [36], [3]. Whilst these works process and classify data points online, classifiers are not updated online. Online classifiers learn incrementally, after each new instance is acquired. The classifier is therefore able to update after every scan, and thus adapt with the data.

During fMRI experiments there may be inherent concept drift. This can be attributed to head motion, physiological changes or low-frequency scanner drift [24]. In addition to this, as stated by LaConte, (2010), future applications of fMRI analysis may consider cases where changes in patterns are *expected* and are a desirable outcome [24]. Experiments involving performance enhancement, rehabilitation or therapy expect the brain response to change over time, with trials being conducted weeks, months or even years apart. In these cases, pretrained classifiers will become less relevant and there is a need for a classifier which adapts and trains over time. LaConte identifies this as one of the current challenges in fMRI classification.

Unfortunately, in many cases, labelled data is not available beyond an initial training phase. Semi-supervised learning provides methods where unlabelled data may be used to update and improve the classifier [29], [34]. One solution, termed naive labelling, uses the predicted label of the incoming data point for updates. In our previous work we use naive labelling within an ensemble framework for i.i.d. (shuffled) fMRI data [32]. This approach however, should be used with caution. If the original classifier is not sufficiently accurate, the online updates may corrupt the the classifier instead of improving it [1]. We found that with the correct parameter tuning the ensemble approach constrained the potential negative behaviour of a classifier with naive updates. In practise, we do not have the option of bespoke parameter tuning. This raises a question as to what is the best strategy when there may be concept drift *and* unlabelled data.

The situation becomes a catch twenty two. We can either use a fixed pretrained classifier which we know will be inaccurate in cases of concept drift, or naive labelling, which may improve the classifier, but runs the risk of making it worse. In this paper we propose an alternative way to use naive labelling within an ensemble framework for streaming fMRI data. We apply what we term *guided updates*: the ensemble prediction is taken to be the 'true' label and is used to update each member classifier instead if its own predicted label. In a related work, [26] also use ensemble labels to boost accuracy in semi-supervised learning, in an offline co-training approach.

The remainder of the paper is organised as follows. Section 2 describes the algorithms. Section 3 introduces the dataset and experimental protocol. Our results are presented in Section 4. Section 5 concludes the paper.

2   Methods

2.1   Online Linear Discriminant Classifier

For this study we use the online linear discriminant classifier (O-LDC). The O-LDC is an adaptation of the linear discriminant classifier, as described in [21]. It is chosen here because, in agreement with common wisdom [12], it was found to be robust and accurate compared to other classifiers for online supervised fMRI classification [31].

2.2   Naive Labelling

In the absence of known class labels we are forced either to use naive (predicted) labels for updates, or to rely on a fixed, pre-trained classifier. A classifier which has been trained solely on a small offline data set is likely to be inaccurate. In addition to this, any concept drift will render a fixed classifier useless.

Training a classifier with naive labelling does not come without risk. The classifier may be led astray should updates occur using incorrectly predicted class labels. This may lead to 'run-away' behaviour where the classifier becomes less accurate as training progresses [1]. The likelihood of runaway classifiers is related to the amount of offline training data and on how well the underlying data distribution model is guessed when designing the classifier [18]. It is expected that the lower the amount of training data, the higher the chances of a runaway classifier.

2.3   Random Subspace Ensemble

In general, classifier ensembles are less sensitive to noise and redundant features than an individual classifier. Over-fitting is therefore less prevalent in classifier ensembles. The Random Subspace ensemble (RS) is a classifier ensemble method whereby ensemble members are trained on feature subsets rather than on the entire feature set [14]. The ensemble decision is based on majority voting. By using RS ensembles the dimensionality of the feature set is reduced, whilst retaining the number of training instances. This makes RS ensembles particularly suitable for datasets with a large feature-to-instance ratio.

A good ensemble should be made up of *diverse* classifiers. The RS method generates diversity by training each ensemble member on a different feature subset. Define $\mathbf{X} = [x_1, \ldots, x_n]^T$ to be the set of n features. To create an RS ensemble, we randomly select L feature subsets of size M by drawing without replacement from a uniform distribution over $\mathbf{X}$. Each subset makes up the feature set for one of the L classifiers. Each of the L classifiers are trained and tested using the respective M features. Ensemble decisions are made by majority vote.

For full-brain fMRI data, the number of features may reach in excess of $80,000$. With appropriate choices of L and M, the RS algorithm is computationally less expensive than ensembles which train on all features, or even a single classifier trained on all the whole dataset. RS ensembles have been shown to perform well for offline fMRI data [22] [23]. In [31] we used the RS ensemble for online classification of labelled fMRI data. [32] shows naive labelling in an ensemble framework to perform favourably against a fixed ensemble, for i.i.d. (shuffled) fMRI data, provided sufficient offline training data was available. This study builds upon our previous work with a modified version of the online naive update strategy for an ensemble framework, results are shown on streaming fMRI data.

2.4   Guided Update Strategy

Classifier ensembles are deemed to be more accurate than individual classifiers [19]. Intuitively, as a consequence, the ensemble predicted label is likely to be more accurate than the predicted label from an individual ensemle member. We hypothesize that by using the ensemble decision to update the individual ensemble members we can reduce the likelihood of runaway classifiers. We expect to see the ensemble with 'guided' updates perform better than an ensemble where its members are updated using their individual predicted labels.

2.5   Theory and Simulations

Before applying the strategy to streaming fMRI data, we first provide the theory for the i.i.d. case.

Consider an ensemble of L classifiers. The ensemble receives a sequence of N i.i.d. data points whose class labels are unknown. If the classifiers in the ensemble are not updated throughout the online run, the ensemble at data point N will be equally accurate as the starting ensemble. Updating the classifiers can improve ensemble accuracy.

Two update strategies can be employed, both within the naive labelling approach.

- Naive (Individual) Update. Each classifier is updated using the label proposed by *the classifier* as the true label.
- Guided (Ensemble) Update. Each classifier is updated using the label proposed by *the ensemble* as the true label.

The naive update can be regarded as a Markov chain where each processed data point is a step in the chain. Denote the initial accuracy of a classifier by p. Assume that, if a correct label is used for the update, the accuracy increases to $p + \epsilon$, if an incorrect label is used, the accuracy decreases to $p - \epsilon$, where ϵ is a small positive constant. The transition matrix for the update step is

		After the update	
Before the update		wrong	correct
	wrong	$1 - p_t + \epsilon$	$p_t - \epsilon$
	correct	$1 - p_t - \epsilon$	$p_t + \epsilon$

Note that the accuracy is tagged by t, the time step. The transition matrix contains the current accuracy p_t which varies from one step to the next. Thus the Markov chain is non-homogeneous, and asymptotic distributions are not readily available.

The probability for correct classification at step $t+1$ can be calculated from the transition matrix

$$p_{t+1} = p_t(p_t + \epsilon) + (1 - p_t)(p_t - \epsilon) = p_t + \epsilon(2p_t - 1). \qquad (1)$$

If the classifier is better than chance at the start ($p > 0.5$), the accuracy is expected to increase progressively with t. For the naive update method, the majority vote accuracy does not play a role in the update. Assuming independent classifiers, the majority vote accuracy will increase with the increase of p.

For the guided update, the probability for correct classification *of each individual classifier* at step $t+1$ depends on the ensemble accuracy, P_{ens} in addition to p_t

$$p_{t+1} = P_{ens}(p_t + \epsilon) + (1 - P_{ens})(p_t - \epsilon) = p_t + \epsilon(2P_{ens} - 1). \qquad (2)$$

Since for independent individual classifiers $P_{ens} > p_t$, the improvement in the individual classifier accuracy will benefit the ensemble updates. The assumptions of i.i.d. data, independent classifiers and that updates lead to improvement (however small) if the correct label is used, cannot be guaranteed in practice.

Sections 3 and 4 introduce the application of this theory to streaming fMRI data, and present our results for this non-i.i.d. case.

3   Experiment with fMRI Data

3.1   Dataset: Bangor 1

Bangor 1 is an fMRI dataset from the School of Psychology, Bangor University. The experiment is of block design. Participants were tasked to invoke negative emotion for blocks of 20s using negative emotional imagery. Blocks of activity were alternated with blocks of rest, lasting 14s each. There were 12 blocks of emotion and 12 blocks of rest. Scans were taken every 2s resulting in 204 brain volumes, which in sequence, represent the change in brain activity over the course of the experiment. The classification task is to distinguish between emotion and rest. Each brain volume represents an instance, voxels within the volume make up the feature set. The true class labels for the instances are taken to be consistent with the stimulus being presented. This is line with one of the protocols suggested by Pereira et al. [30]. This does not take into account the Haemodynamic Reponse Function (HRF), however it is noted that the HRF varies between subjects, psychological states, experimental conditions and different brain regions [11]. Also, as the transition between brain states is gradual, it is unclear at which point we should say one state ceases and the next begins. Assigning labels according to the stimuli standardises the experiment.

Data was collected on a 3 Tesla Philips Achieva MR scanner (TR = 2 s, TE = 30 ms, 30 slices, in-plane resolution $2 \times 2\text{mm}^2$, 3 mm slice thickness). Pre-processing of the data was performed using Brainvoyager QX (Braininnovation, Maastricht, The Netherlands). The data were corrected for intra-subject angular and translational motion and filtered to remove long-term drift [17].

3.2 Protocol

To simulate a real time scenario, the first 17 instances are taken to form an offline training sample, T. These instances represent the first block of emotion and first block of rest in the experiment. Following the recommended procedure by De Martino et al. [2] in their work on feature selection, we use a univariate method to pre-select $K = 2000$ features. An ANOVA test is used on T for feature selection. The p-value of this test is used to rank the voxels. A subset of the K voxels with the lowest p-values are chosen as features for the classification. The remaining 187 instances, in sequence, make up the online training data, S.

An RS ensemble is trained on T using the algorithm described in Section 2.3. The same base ensemble is used for each of the update strategies. Instances from S are presented sequentially, with the following procedure being applied:

- **Fixed strategy**
Ensemble accuracy is tested on the instance. No update is carried out.
- **Naive strategy**
Ensemble accuracy is tested on the instance. Ensemble is updated using predicted labels from the *individual* classifiers.
- **Guided strategy**
Ensemble accuracy is tested on the instance. Ensemble is updated using the *ensemble* decision.

We repeat the experiment for parameters $L = [5, 9, 13]$, $M = [20, 50, 100, 250]$.

4 Results

For each time-step t the cumulative error is calculated as $\frac{\sum_{j=0}^{t} e(j)}{n}$, where $e(j)$ is 0, if the ensemble has labelled the point at time j correctly, and 1, otherwise. Figure 1 plots of the cumulative error scores over time for parameters $L = 13$ and $M = 20$. The vertical lines indicate the class boundaries. The plot is taken from $t = 25$, for lower values of t the plot appears noisy due to the calculation of the cumulative error. Residual noise in the plot can be seen by the apparent decline in error of the fixed ensemble up to $t = 60$. From this point, the error of the fixed ensemble stabilises at 22.5%, the error of the other ensembles continues to decline showing that the ensembles learn and adapt with the data.

The presence of a sequence of multiple instances from the same class (due to the block design of the experiment) can be seen to affect the ensembles in that with every class change a small peak is seen in the error level. This peak arises where the ensemble sees data points from the 'transition' period, where the true

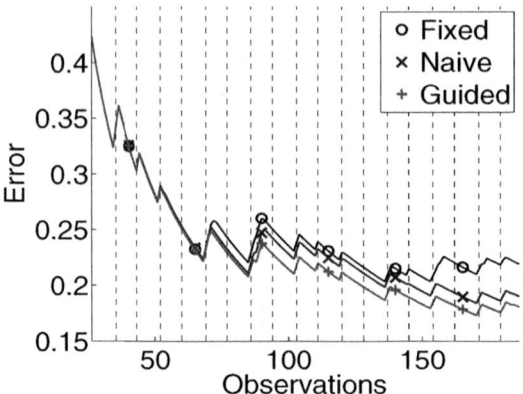

Fig. 1. Cumulative error progression comparing the three strategies for $L = 13$, $M = 20$. Vertical lines represent class boundaries.

state of the brain is uncertain. The transition is the period when we expect the ensemble to make most mistakes.

The 'final' error scores for all parameter combinations, taken at $t = 187$, are given in Table 1.

Table 1. Error (%) 'F', 'N' and 'G' correspond to Fixed, Naive and Guided strategies respectively

	$M = 20$			$M = 50$			$M = 100$			$M = 250$		
	F	N	G	F	N	G	F	N	G	F	N	G
$L = 5$	22.2	19.6	18.35	21.7	18.9	18.2	22.0	19.6	19.5	22.0	20.2	21.22
$L = 9$	22.0	18.9	17.8	21.9	18.6	18.1	22.0	18.9	19.4	22.0	20.9	22.1
$L = 13$	21.8	18.9	17.95	21.9	18.4	17.8	22.0	19.1	19.4	22.0	20.6	21.9

Direct comparison of the naive and guided strategies with the fixed ensemble is offered in Table 2. Strategies which perform better than the fixed classifier are indicated by a '+', strategies which perform worse are indicated by a '−'. Significant results, tested using a paired t-test with significance $\alpha = 0.05$, are indicated by $\oplus$ and $\ominus$ respectively. Both the naive and guided ensembles can be seen to perform significantly better than the fixed ensemble for the vast majority of parameters.

For each strategy and parameter set, strategies are tested in a pairwise manner. Results are tested for significance using a paired t-test (at significance $\alpha = 0.05$). We compare both the final error scores and the average error. The average error corresponds to the area under the curve and thus gives an indication of the learning capability of the strategy. As we have 12 parameter sets and 3 strategies, a total of 36 pairwise comparisons are made. The numbers of wins versus losses are plotted in Figure 2. The best point is at 24 wins and 0 losses,

Table 2. Direct comparison of the naive and guided strategies with the fixed ensemble. Strategies which perform better than the fixed classifier are indicated by a '+', strategies which perform worse are indicated by a '−'. Significant results are indicated by ⊕ and ⊖ respectively.

	$L = 5$		$L = 9$		$L = 13$	
	Naive	Guided	Naive	Guided	Naive	Guided
$M = 20$	⊕	⊕	⊕	⊕	⊕	⊕
$M = 50$	⊕	⊕	⊕	⊕	⊕	⊕
$M = 100$	⊕	⊕	⊕	⊕	⊕	⊕
$M = 250$	⊕	⊕	⊕	−	⊕	+

the worst point at 0 wins and 24 losses. The fixed strategy is seen to have the worst results in each plot. The guided strategy performs best, albeit by a small amount. This indicates that using the ensemble decision to update the classifiers is beneficial. Making use of the higher accuracy of the ensemble decision constrains potential runaway behaviour in individual ensemble members, which in turn leads to a more accurate ensemble.

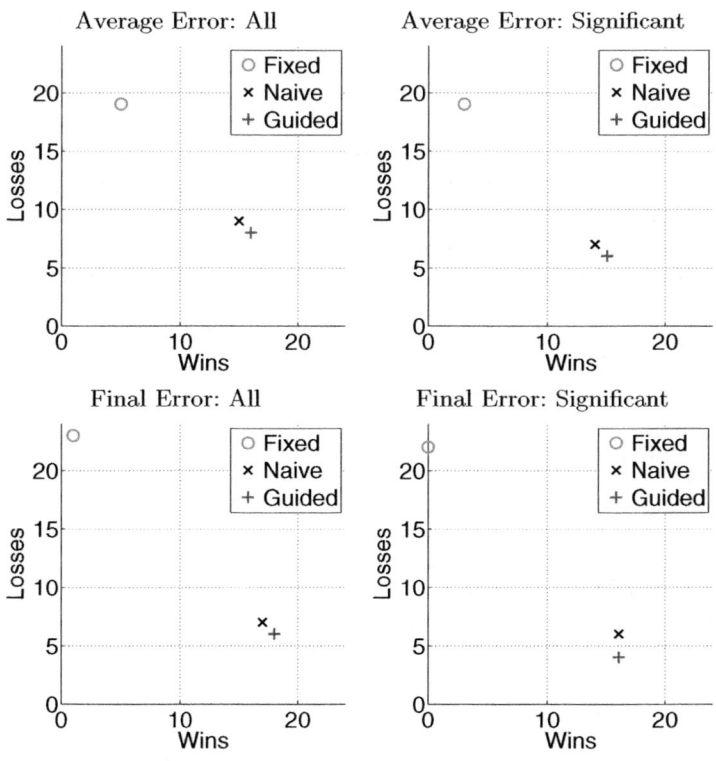

Fig. 2. Pairwise wins vs losses. Significance calculated at $\alpha = 0.05$.

5   Conclusion

Real-time fMRI classification faces challenges of unlabelled data and concept drift. This study proposes a solution in the form of classifier ensembles. The solutions have been tested and illustrated on streaming fMRI data. The experiments show that the ensembles benefit from updating during the online phase. Both proposed update strategies are shown to perform significantly better than the fixed strategy across a variety of parameters.

The guided update strategy offers a possible solution to the combination of unlabelled data and concept drift. Results from the update strategy compare well with the standard naive classifier ensemble, and perform significantly better than the fixed classifier ensemble. The guided ensemble has the lowest error score of the three ensembles tested in seven out of twelve parameter combinations.

Future work may apply the strategies across a wider range of parameters, and on more datasets, in particular those with known concept drift. Other directions include combining the strategy with different base classifiers or a different ensemble framework or voting strategy.

Acknowledgments. We would like to thank Dr. D. Linden and Dr. S. Johnston from the School of Psychology, Bangor University, for supplying us with the Bangor 1 dataset.

References

1. Cozman, F.G., Cohen, I.: Unlabeled Data can Degrade Classification Performance of Generative Classifiers. In: Proceedings of the 15th International FLAIR Conference, pp. 327–331 (2002)
2. De Martino, F., Valente, G., Staeren, N., Ashburner, J., Goebel, R., Formisano, E.: Combining multivariate voxel selection and support vector machines for mapping and classification of fMRI spatial patterns. NeuroImage 43(1), 44–58 (2008)
3. de Charms, R.C.: Applications of real-time fMRI. Nature Reviews Neuroscience 9(9), 720–729 (2008)
4. de Charms, R.C., Maeda, F., Glover, G.H., Ludlow, D., Pauly, J.M., Soneji, D., Gabrieli, J.D.E., Mackey, S.C.: Control over brain activation and pain learned by using real-time functional MRI. Proc. Natl. Acad. Sci. USA 102(51), 18626–18631 (2005)
5. Eklund, A., Andersson, M., Ohlsson, H., Ynnerman, A., Knutsson, H.: A Brain Computer Interface for Communication Using Real-Time fMRI. In: International Conference on Pattern Recognition (2010)
6. Eklund, A., Ohlsson, H., Andersson, M., Rydell, J., Ynnerman, A., Knutsson, H.: Using Real-Time fMRI to Control a Dynamical System by Brain Activity Classification. In: Yang, G.-Z., Hawkes, D., Rueckert, D., Noble, A., Taylor, C. (eds.) MICCAI 2009. LNCS, vol. 5761, pp. 1000–1008. Springer, Heidelberg (2009)
7. Fleiss, J.L.: Statistical Methods for Rates and Proportions. John Wiley & Sons (1981)
8. van Gerven, M., Farquhar, J., Schaefer, R., Vlek, R., Geuze, J., Nijholt, A., Ramsay, N., Haselager, P., Vuurpijl, L., Gielen, S., Desain, P.: The Brain-Computer Interface Cycle. Journal of Neural Engineering 6 (2009)

9. Golub, T.R., Slonim, D.K., Tamayo, P., Huard, C., Gaasenbeek, M., Mesirov, J.P., Coller, H., Loh, M.L., Downing, J.R., Caligiuri, M.A., Bloomfield, C.D., Lander, E.S.: Molecular Classification of Cancer: Class Discovery and Class Prediction by Gene Expression Monitoring. Science 286(5439), 531–537 (1999)
10. Guyon, I., Hur, A.B., Gunn, S., Dror, G.: Result analysis of the NIPS 2003 feature selection challenge. In: Advances in Neural Information Processing Systems, vol. 17, pp. 545–552 (2004)
11. Handwerker, D.A., Ollinger, J.M., D'Esposito, M.: Variation of BOLD hemodynamic responses across subjects and brain regions and their effects on statistical analyses. Neuroimage 21(4), 1639–1651 (2004)
12. Hastie, T., Tibshirani, R., Friedman, J.: The Elements of Statistical Learning. Springer, Heidelberg (2001)
13. Haxby, J.V., Gobbini, M.I., Furey, M.L., Ishai, A., Schouten, J.L., Pietrini, P.: Distributed and overlapping representations of faces and objects in ventral temporal cortex. Science 293(5539), 2425–2430 (2001)
14. Ho, T.K.: The random space method for constructing decision forests. IEEE Transactions on Pattern Analysis and Machine Intelligence 20(8), 832–844 (1998)
15. Holmes, A.J., MacDonald, A., Carter, C.S., Barch, D.M., Stenger, V.A., Cohen, J.D.: Prefrontal functioning during context processing in schizophrenia and major depression: An event-related fMRI study. Schizophrenia Research 76, 199–206 (2005)
16. Hugdah, K., Rund, B.R., Lund, A., Asbjornsen, A., Egeland, J., Ersland, L., Landr, N.I., Roness, A., Stordal, K.I., Sundet, K., Thomsen, T.: Brain Activation Measured With fMRI During a Mental Arithmetic Task in Schizophrenia and Major Depression. American Journal of Psychiatry 161, 286–293 (2004)
17. Johnston, S.J., Boehm, S.G., Healy, D., Goebel, R., Linden, D.E.J.: Neurofeedback: A promising tool for the self-regulation of emotion networks. Neuroimage 29 (2009)
18. Kuncheva, L., Whitaker, C., Narasimhamurthy, A.: A case study on naïve labelling for the nearest mean and the linear discriminant classifiers. Pattern Recognition 41, 3010–3020 (2008)
19. Kuncheva, L.I.: Combining Pattern Classifiers: Methods and Algorithms. Wiley Interscience (2004)
20. Kuncheva, L.I., Plumpton, C.O.: Choosing parameters for random subspace ensembles for fMRI classification. In: Proc. Multiple Classifier Systems (2010)
21. Kuncheva, L.I., Plumpton, C.O.: Adaptive learning rate for online linear discriminant classifiers. In: Proc. S+SSPR, Orlando, Florida, USA, pp. 510–519 (2008)
22. Kuncheva, L.I., Rodríguez, J.J.: Classifier ensembles for fMRI data analysis: An experiment. Magnetic Resonance Imaging 28, 583–593 (2010)
23. Kuncheva, L.I., Rodríguez, J.J., Plumpton, C.O., Linden, D.E.J., Johnston, S.J.: Random subspace ensembles for fMRI classification. IEEE Transaction on Medical Imaging
24. LaConte, S.M.: Decoding fMRI brain states in real-time. NeuroImage (2010)
25. Lang, P., Bradley, M., Cuthbert, B.: International Affective Picture System (IAPS): Technical Manual and Affective Ratings
26. Li, M., Zhou, Z.-H.: Improve computer-aided diagnosis with machine learning techniques using undiagnosed samples. IEEE Transactions on Systems, Man and Cybernetics A 37(6), 1088–1098 (2007)
27. Misaki, M., Kim, Y., Bandettini, P.A., Kriegeskorte, N.: Comparison of multivariate classifiers and response normalizations for pattern-information fMRI. NeuroImage (2010), doi:10.1016/j.neuroimage.2010.05.051

28. Moench, T., Hollmann, M., Grzeschik, R., Muller, C., Luetzkendorf, R., Baecke, S., Luchtmann, M., Wagegg, D., Bernarding, J.: Real-time classification of activated brain areas for fMRI-based human-brain-interfaces. In: Medical Imaging 2008: Physiology, Function, and Structure from Medical Images, vol. 6916, pp. 69161R – 69161R-10 (2008)
29. Nigam, K.P.: Using Unlabeled Data to Improve Text Classification. PhD thesis, School of Computer Science, Carnegie Mellon University, Pittsburgh, US (2001)
30. Pereira, F., Mitchell, T., Botvinick, M.: Machine learning classifiers and fMRI: a tutorial overview. NeuroImage 45, S199–S209 (2009)
31. Plumpton, C.O., Kuncheva, L.I., Linden, D.E.J., Johnston, S.J.: On-line fMRI Data Classification Using Linear and Ensemble Classifiers. In: Proc. 20th International Conference on Pattern Recognition (2010)
32. Plumpton, C.O., Kuncheva, L.I., Oosterhof, N.N., Johnston, S.J.: Naive random subspace ensemble with linear classifiers for real-time classification of fMRI data. Pattern Recognition special edition on Brain Decoding (in Press, Corrected Proof, 2011)
33. Posse, S., Fitzgerald, D., Gao, K., Habel, U., Rosenberg, D., Moore, G.J., Schneider, F.: Real-time fMRI of temporolimbic regions detects amygdala activation during single-trial self-induced sadness. NeuroImage 18, 760–768 (2003)
34. Seeger, M.: Learning with labeled and unlabeled data. Technical report, University of Edinburgh (2001)
35. Sheline, Y.I., Barch, D.M., Donnelly, J.M., Ollinger, J.M., Snyder, A.Z., Mintun, M.A.: Increased Amygdala Response to Masked Emotional Faces in Depressed Subjects Resolves with Antidepressant Treatment: An fMRI Study. Biological Psychiatry 50(9), 651–658 (2001)
36. Weiskopf, N., Sitaram, R., Josephs, O., Veit, R., Scharnowski, F., Goebel, R., Birbaumer, N., Deichmann, R., Mathiak, K.: Real-time functional magnetic resonance imaging: methods and applications. Magnetic Resonance Imaging 25, 989–1003 (2007)
37. Yoo, S.S., Fairneny, T., Chen, N.K., Choo, S.E., Panych, L.P., Park, H.W., Lee, S.Y., Jolesz, F.A.: Brain–computer interface using fMRI: spatial navigation by thoughts. NeuroReport 15(10), 1591–1595 (2004)

Studying Self- and Active-Training Methods for Multi-feature Set Emotion Recognition

José Esparza[1], Stefan Scherer[2], and Friedhelm Schwenker[1]

[1] Institute of Neural Information Processing
University of Ulm, Germany
[2] School of Linguistic, Speech and Communication Sciences
Trinity College Dublin, Ireland

Abstract. Automatic emotion classification is a task that has been subject of study from very different approaches. Previous research proves that similar performance to humans can be achieved by adequate combination of modalities and features. Nevertheless, large amounts of training data seem necessary to reach a similar level of accurate automatic classification. The labelling of training, validation and test sets is generally a difficult and time consuming task that restricts the experiments. Therefore, in this work we aim at studying self and active training methods and their performance in the task of emotion classification from speech data to reduce annotation costs. The results are compared, using confusion matrices, with the human perception capabilities and supervised training experiments, yielding similar accuracies.

Keywords: Human perception of emotion, automatic emotion classification, semi-supervised learning, active learning, emotion recognition from speech.

1 Introduction

Emotion classification relies, as all classification problems, in the features that support it and their variability for the different classes considered. Literature shows that in the case of emotion classification, there exist many situations where not even an expert - human - is capable of emitting a decision with absolute confidence, due to real overlappings between the different classes. For these scenarios, where cross-class confusions are unavoidable in some cases, large training sets are often required in order to achieve accurate enough results.

Previous research aimed at emulating human perception capabilities shows that by means of choosing appropriate feature sets and exhaustive training, similar accuracies and confusions may be obtained by using large training sets. This, however, implies a tedious labelling process conducted by experts which, in general, may represent a very expensive and time consuming effort. Further, not all manual annotations might improve the automatic classifier's performance, as uninformative data (e.g. data far from decision boundaries) hardly influences the discriminative performance of the classifier.

F. Schwenker and E. Trentin (Eds.): PSL 2011, LNAI 7081, pp. 19–31, 2012.
© Springer-Verlag Berlin Heidelberg 2012

Obtaining unlabelled samples, however, does not necessarily incur in high costs and large amounts of data should be exploitable even without annotations. For this reason, there is continuous research being conducted with the aim of using unlabelled data for training. To make use of this unlabeled training data, different approaches and research lines, each of them focusing on different properties of the training process, exist. There is research conducted, for example, on semi-supervised learning, where both labelled and unlabelled data are used for model training ([3], [20], [2]), unsupervised learning, where only unlabelled data is used (eg. Clustering algorithms - [4]) or active learning, where the system is allowed to choose its training data from a pool of samples [8].

In this work we use both a semi-supervised approach based on k-nearest neighbor algorithm providing preliminary fuzzy estimates and an active learning approach for training multi-classifier multi-class support vector machines (SVM). Eight separate feature sets extracted from speech data are combined to assess the performance on a standard emotion dataset.

The remainder of the paper is organized as follows: Section 3 introduces the used datasets and the human perception benchmarks, reported as confusion matrices. Section 4 then describes the employed feature sets, as well as the encoding of sequential features. The experimental setup is briefly described in Section 2 and the results are reported in Section 5. The automatic classification performances are then compared with the human perception in Section 6, and Section 7 concludes the paper.

2 Methods

The experiments were conducted on the full WaSeP dataset with six target categories. The gender-independent experiment was conducted with data from both male and female speakers. For each feature set and class, two hidden Markov models (HMM) are trained with male-only data and another 2 with female-only data. To train the SVM, also equal amount of data from male and female speakers was used. Results were calculated without considering whether the test samples were produced by a male or female speaker. In the following the three separate experimental setups are introduced briefly.

2.1 Supervised Learning Experiment

For the supervised learning, which serves as a benchmark for the latter experiments, we utilized the F^2SVM introduced in [16]. For each feature set an F^2SVM was trained separately. The different fuzzy outputs of each SVM are combined by a simple multiplication fusion and normalization. A ten fold cross validation with a 90% training and 10% test data-set split for the evaluation was conducted.

2.2 Self-training Experiment

In this experiment, we would like to aim at automatically generate fuzzy labels for unseen data, starting from a small reference set, for which labels are available, before training the same F^{2}SVM architecture as in Section 2.1. Although there exist different techniques for self-training, only k-nearest neighbour (k-NN) is utilized in this work, with $k = 5$. For each unlabeled point, a new fuzzy label is generated by averaging the labels of the closest k reference points. The newly generated label is then included into the reference set and considered as correct for all the still unlabeled samples, thereby the reference set increases iteratively. The iterations are repeated for all the unseen data-points. When the new fuzzy labels are generated, the SVMs are trained in a supervised style, assuming that the automatically generated labels are correct, leading to a semi-supervised approach. In order to control the amount of error introduced by the automatically generated labels, they are processed discrimination process with a pivot parameter p. Labels with a confidence higher than p are used for training the SVMs and those with a lower confidence are discarded. A graphical representation of the process and the training set selection can be seen in Figures 1 and 2 respectively.

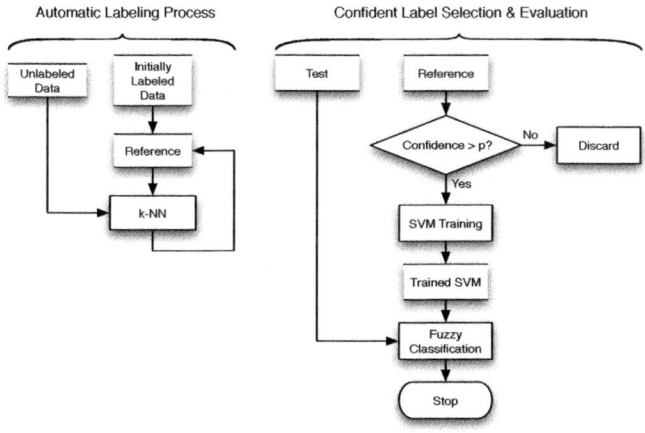

Fig. 1. k-NN Flow Chart

2.3   Active Learning Experiment

Traditional machine learning approaches rely on a large amount of labelled data distributed over the feature spaces with as much information as possible concerning the underlying generative distribution. These experiments are aimed at reducing the required amount of training data by letting the system choose the samples itself. The most striking research question here is of course the choice and selection of the most relevant samples that could improve the performance.

First of all, the whole available dataset with available labels is divided in two groups (i.e. training and test[1]). The training set, represents the pool of available

---

[1] Note that the test set remains unchanged during the whole process.

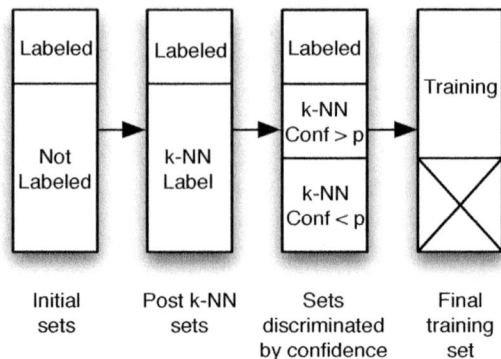

Fig. 2. k-NN Training sets

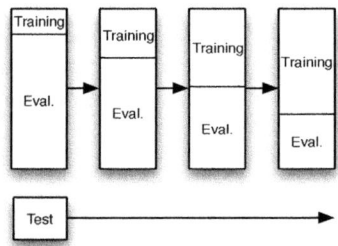

Fig. 3. Active learning training-evaluation-test sets evolution

data from which the system will decide on every iteration which labels it wants to have labels for and use for training. The evolution of the sets over the iterations is represented in Figure 3.

A small number of labels is initially used for the training, then evaluation is conducted on the unused training points. For each of these points, the SVMs produce a fuzzy output label that represents the degree of membership to all the classes. The accumulated membership to the classes must be equal to 1 and, therefore, considering the highest membership in one label also accounts for the most likely class. It is then possible to define the confidence of the label as the degree of membership to the most likely class.

Considering only the most likely class for each label can be assumed to provide a measure of how confident a decision is. In this case the more confident a decision might be the less relevant for improvement it might be. Under this assumption, it makes sense to believe that low confidence labels are the ones that the system has trouble in classifying. Further, while considering the architecture of the utilized classifiers, i.e. F^2SVM, low values indicate proximity to the decision boundary. Therefore, these samples might be the most informative influencing the decision boundary in further iterations. On the other hand, for the output labels that

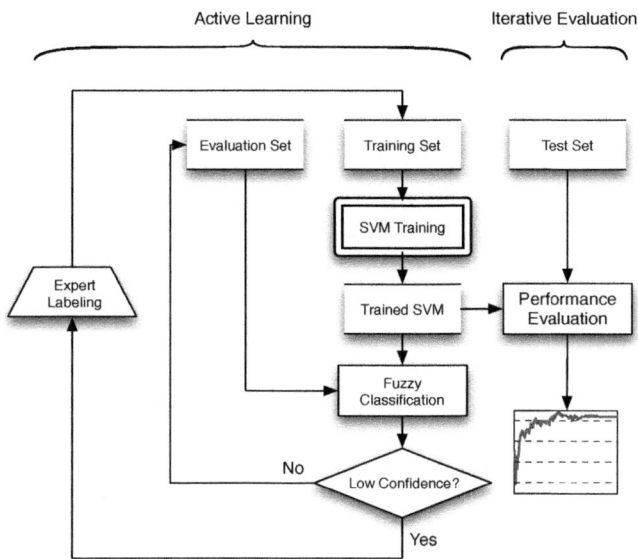

Fig. 4. Flow Chart of the active learning and evaluation process

show a good confidence, the system does not require more information since they represent an easier task for it. A flow chart representing the whole learning and evaluation process is presented in Figure 4.

3   Dataset Description

The experiments in this work are based on the "Corpus of spoken words for studies of auditory speech and emotional prosody processing" (WaSeP©) [19], which consists of two main parts: a collection of German nouns and a collection of phonetically balanced pseudo words, which correspond to the phonetical rules of German language, such as "hebof", "kebil", or "sepau". For this study the pseudo words have been chosen as the basis. This pseudo word set consists of 222 words, repeatedly uttered by a male and a female actor in six different emotional prosodies: neutral, joy, sadness, anger, fear, and disgust. The average duration of the speech signals depends on the specific emotion, ranging from .75 sec. in the case of the "neutral" prosody, to 1.70 sec. in the case of "disgust". The data was recorded using a Sony TCD-D7 DAT-recorder and the Sennheiser MD 425 microphone in an acoustic chamber with a 44.1 kHz sample rate and later down-sampled to 16 kHz with a 16 bit resolution. Furthermore, a perception test has been conducted with 74 native German listeners, who were asked to rate and name the category or prosody that they were just listening to, resulting in an overall accuracy of 78.53%. Table 1 shows the confusion matrix of the human perception test. It was also observed that the most confused emotion is "disgust", which is conform with the assumptions of [12].

Table 1. Confusion matrix of the human performance test generated from the available labels for each of the utterances listed in the WaSeP database, [18]

	F	D	H	N	S	A
Fear	**0.77**	0.01	0.08	0.03	0.10	0.01
Disgust	0.05	**0.72**	0.06	0.03	0.07	0.07
Happiness	0.01	0.00	**0.75**	0.22	0.02	0.00
Neutral	0.01	0.02	0.05	**0.79**	0.00	0.13
Sadness	0.05	0.01	0.04	0.13	**0.76**	0.01
Anger	0.01	0.03	0.00	0.01	0.01	**0.94**

4   Features

In similar work, different combinations of audio features are said to perform well in classification of emotional audio data [7]. Given the characteristics of the used data set, the chosen features for this work are the following:

1. *MFCC / ΔMFCC*: based on the human perceptual scale of pitches. For the *MFCC* extraction a window length of 25 ms and a shift time of 10 ms is used, with a total of 20 cepstral coefficients, as well as their derivatives [11].
2. *modSpec*: implemented in an attempt to measure the *modulation* of the spectral coefficients. This is a way of accounting how much and how fast the features vary over time [9,5].
3. *Voice Quality*: the dynamic use of voice qualities in spoken language can reveal useful information on a speakers attitude, mood and affective states. The exact set of the utilized features is described in detail in [13].
4. f_0: it is possible to obtain different values of f_0 over time. From the f_0 trail different statistics are calculated: mean, standard deviation, maximum and quartile values, forming the feature set.
5. *Energy*: the frame average energy is calculated using a window size of 32 *ms* with an overlap of 16 *ms*. Similar statistics to those of f_0 are used for this.
6. *PLP*: perceptual linear predictive (PLP) analysis is based on perceptually and biologically motivated concepts, the critical bands, and the equal loudness curves, as described in [6].
7. *Periodicity*: This set is designed based on correlation measures of the speech signals. From the idea that vowels have a higher periodicity than consonants, this measures can be considered as an indicator of the syllables speed. For this purpose, different statistics from the relation of periodic segments over the total length are used as feature. Similar parameters are also obtained from the energy distribution.

Emotion classification from speech data proves to be a challenging problem due to the sequential nature of the data. Therefore, dynamic features extracted on short segments of speech (32ms windows) are useful for the classification of expressive clips. However, in order to be able to compare and combine these sequential features in a multi-classifier architecture with static features it is necessary to encode them into vectors of a fixed length. There exist different approaches for dealing with this type of situations. In this work, vectorial HMM, as in [1], are used to encode the sequential data to a new representation space, where every sequence can be represented in terms of a fixed number of dimensions.

Additionally, since the feature spaces are usually very heterogeneous, data normalization is performed. During the training of the system, mean and standard deviation (μ_{train} and σ_{train}) are calculated in each feature domain and for each class, prior to the HMM training. To remove the effect of outliers, all values above and below the 95% and 5% percentiles, respectively, are discarded. With the normalized data, the HMM are trained and the same normalization values (μ_{train} and σ_{train}) are later used to normalize the unseen data in the test step, before calculating their likelihood values.

5   Experiments

Confusion matrices have been computed to analyse decisions. Every row sums up to one, showing how much data from one class is classified by the system as belonging to any of the possible ones. The columns (which do not necessarily sum up to one) show how much data from all classes is classified as part of a given one. Results for supervised learning experiment results are also included for comparison with the partially-supervised experiments performance.

5.1   Supervised Learning

The classification accuracy in the gender-independent test is resulted in an average accuracy of 84%. Happiness produces the lowest number of hits, being highly confused with fear and neutral. A paired t-test shows a highly statistically significant improvement for the fusion over the single best feature set, namely MFCC ($p < .001$). For example, in the case of disgust or happiness (the categories with the lowest accuracy), an increase of .08 in F_1 measure can be achieved. The confusion matrix is shown in Table 2.

Table 2. Confusion matrix of fused features for the gender-independent automatic classification experiments, conducted with the WaSeP dataset

	F	D	H	N	S	A
Fear	**0.80**	0.03	0.08	0.01	0.02	0.06
Disgust	0.01	**0.88**	0.05	0.00	0.04	0.03
Happiness	0.08	0.02	**0.71**	0.12	0.04	0.03
Neutral	0.00	0.01	0.16	**0.82**	0.01	0.00
Sadness	0.02	0.00	0.03	0.01	**0.95**	0.00
Anger	0.01	0.07	0.02	0.03	0.00	**0.86**

5.2   Self-training Experiment

Several experiments have been conducted within this approach with the aim to produce a significant improvement in the classification performance when the

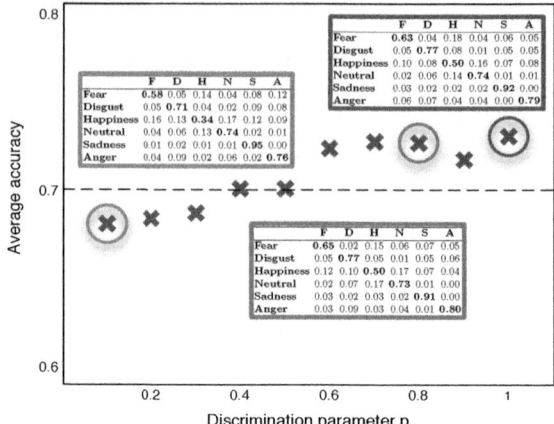

Fig. 5. Average accuracy obtained for different values of the discrimination parameter p. The confusion matrices obtained for values of p 0.1 and 0.8 are shown. As well as these, the confusion matrix that represent the baseline for this experiment is obtained for a value of p equal to 1, since this is the highest value of the discrimination parameter and only real labels are able of reaching it.

system is trained with a reduced set of crisp labels, extended with a large number of fuzzy automatically generated labels.

The baseline in this experiment has been lowered to resemble a situation with small amounts of data available. This baseline provides an average accuracy of 73% for the gender-independent case, with only 20 samples per emotional category available. A sweep analysis over the parameter p shows that the maximum is found for a discrimination value of $p = 0.8$, achieving also an average of 73%. Graphical representation of this analysis is shown in Figure 5, where gender-independent results obtained are also shown as confusion matrices for values of the discrimination parameter p equal to 0.1, 0.8 and 1. Since no improvement is observed by extending the SVM training set with automatically labelled samples, it seems logical to believe that either the used confidence measure is not valid or the generated kNN labels contain too much error.

A second experiment has been carried out to check the effect of the error introduced into the labels by the k-NN algorithm. For this purpose, a larger reference set was used for generating k-NN labels, but not completely used for training the SVMs, as described in Figure 6. In this way it is possible to observe the effect of the discriminative parameter p over the system accuracy, as shown in Figure 7.

5.3   Active Learning Experiment

In the iterative training and evaluation process, each iteration represents an increase of 10 samples in the training set. For evaluation of the results obtained in

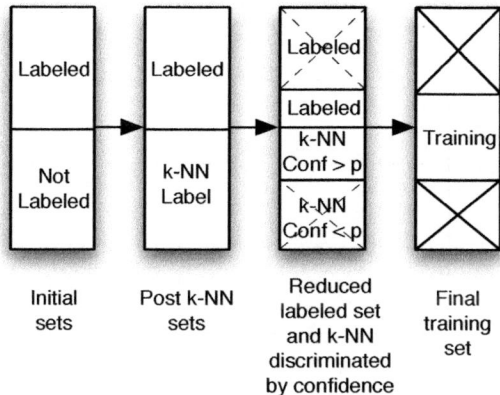

Fig. 6. Evolution of the training sets over time for validation of the confidence measure validity. k-NN labels are automatically generated from an extended reference set to reduce the error in them. To train the SVMs, not all the reference set is used, but only a small fraction of it, together with the artificially labeled samples of confidence higher than the discrimination parameter p.

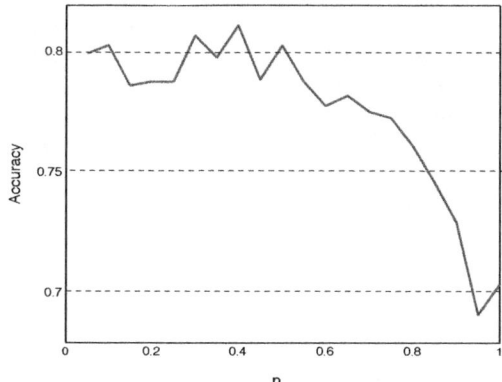

Fig. 7. Classification accuracy for different values of p using k-NN labels generated with a large reference set. The reference set was later reduced and only a 10% of it was used in the SVM training together with the new labels.

this section, Figure 8 has been generated. This figure shows the average accuracy of the trained system for each step of the iterative process. Table 3 shows the confusion matrix of the active learning experiment after the last iteration (with all the available training data used). Average accuracy in this case, 88.2% is higher than the 84% obtained in section 5.1 due to a larger training set utilized for a better representation of the effect produced by the active learning. Further, it should be noted that after only a few iterations some sort of saturation point is reached, that comprises only a small portion of the available data for training.

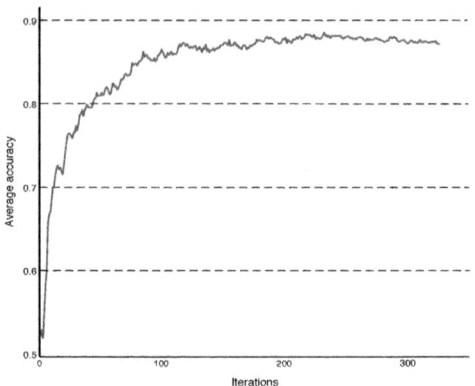

Fig. 8. Active learning accuracy over iterations, for the gender-independent case conducted with the WaSeP dataset. Each iteration represents 10 new labels used for training.

Table 3. Confusion matrix of the gender-independent active learning experiments

	F	D	H	N	S	A
Fear	**0.83**	0.01	0.11	0.01	0.01	0.02
Disgust	0.01	**0.89**	0.05	0.01	0.02	0.03
Happiness	0.05	0.02	**0.79**	0.10	0.02	0.02
Neutral	0.01	0.00	0.10	**0.88**	0.00	0.01
Sadness	0.01	0.00	0.02	0.01	**0.97**	0.00
Anger	0.01	0.02	0.02	0.01	0.00	**0.93**

6   Discussion

The confusion matrix provided in Section 5 provides a good basis for the comparison of human and machine capabilities and errors, as well as the different training approaches under study. A first glance at the numbers shows that human and machine performances are quite similar on an overall scale. With the WaSeP dataset, the 84% accuracy rate obtained is exactly the same as that of humans in average. These figures, however, shall be used to compare the wellness of the experiments conducted within the partially-supervised framework.

As for the semi-supervised experiments with the data labelled by the k-NN algorithm, analysis of the obtained results (see Figure 5) shows that the use of unlabelled data in the training process does not improve the baseline. The baseline in this experiment has been lowered to resemble a situation with small amounts of data available (i.e. only 20 samples per category). This baseline provides an average accuracy of 73% for the gender-independent case. As already commented in section 5.2, a sweep simulation over different values of p was conducted, finding its maximum at $p = 0.8$. This maximum, however, is not an increase with respect to the case where no unlabelled data is considered for training. Given the large amount of automatically generated labels used in the training, it is wise to think that the style in which the experiments were designed was not correct. There might be different reasons for this, like a bad selection of the confidence measure or the excessive amount of error present in the self-labelled samples. Assuming that the chosen confidence measure is correct, better results are to be expected if the artificial labels are more accurately generated. To prove this assumption, a second experiment has been conducted where the aim was to reduce the error artificially put into the k-NN labels. The set of labelled data used for training the SVMs is now reduced in order to be able to measure the accuracy with most of the training data obtained by the k-NN process. As can bee seen in Figure 7, when not a large amount of error is introduced artificially, good performance improvements can be achieved if training with unlabelled data. It makes sense to believe that with an automatically labelling algorithm that inserts less error than k-NN it may be possible to use semi-supervised learning with good accuracy results. As, the utilised confidence measure proved to give good results when the artificial labels contain more correct information.

In opposition to the poor results encountered with the semi-supervised approach, active learning proved to be a very good approach for reducing the amount of labelled data required. It can be seen that after approximately 60 iterations (100 training samples per class) the accuracy already reaches a similar level performance to that of the supervised learning approach, using twice as much data. This means a large reduction of the required amount of labelled data, proving that the approach works and produces good results. In Figure 8 it is observed that after a certain iteration, the addition of new labelled data does not lead to an accuracy increase. We can, therefore, affirm that the active learning works well and can significantly reduce the required amount of data without penalising the obtained results.

7   Conclusions

In the task of emotion classification, there is documented prove that humans perform with higher error rates than in other recognition tasks. In this work, we compared automatic emotion classification with the human performance and studied different partially supervised approaches for training a classifier. In particular, we proved that a semi-supervised approach with artificial labels generated by k-NN does not produce good results due to a large amount of error introduced automatically by the system. In opposition to these, good results were obtained for experiments conducted in an active learning style, where a reduction of the training data with respect to the supervised training case still produces an accuracy comparable to that achieved in the human perception tests. Future work should include the study of more sophisticated self-labelling methods in order to improve the poor self-training results obtained. As for the active training, different approaches to the one utilized exist and should also be studied [15,14,10,17]. Further, as the utilized datasets were composed of acted speech segments, future work should include the study of natural data as well as a deeper knowledge of the representative characteristics of each different emotion.

Acknowledgment. This paper is based on work done within the Transregional Collaborative Reserach Centre SFB/TRR 62 *Companion Technology for Cognitive Technical Systems* funded by the German Research Foundation (DFG).

References

1. Bicego, M., Murino, V., Figueiredo, M.: Similarity-Based Clustering of Sequences using Hidden Markov Models. In: Perner, P., Rosenfeld, A. (eds.) MLDM 2003. LNCS, vol. 2734, pp. 95–104. Springer, Heidelberg (2003)
2. Blum, A., Mitchell, T.: Combining labeled and unlabeled data with co-training. In: Proceedings of the Eleventh Annual Conference on Computational Learning Theory, COLT 1998, pp. 92–100. ACM, New York (1998)
3. Druck, G., Mann, G., McCallum, A.: Learning from labeled features using generalized expectation criteria. In: Proceedings of the 31st Annual International ACM SIGIR Conference on Research and Development in Information Retrieval, SIGIR 2008, pp. 595–602. ACM, New York (2008)
4. Duda, R.O., Hart, P.E., Stork, D.G.: Pattern Classification, 2nd edn. Wiley, New York (2001)
5. Hermansky, H.: The modulation spectrum in automatic recognition of speech. In: Proceedings of IEEE Workshop on Automatic Speech Recognition and Understanding, pp. 140–147. IEEE (1997)
6. Hermansky, H., Morgan, N.: Rasta processing of speech. IEEE Transactions on Speech and Audio Processing, special issue on Robust Speech Recognition 2, 578–589 (1994)
7. Li, D., Sethi, I.K., Dimitrova, N., McGee, T.: Classification of general audio data for content-based retrieval. Pattern Recognition Letters 22(5), 533–544 (2001)
8. Lomasky, R., Brodley, C.E., Aernecke, M., Walt, D., Friedl, M.: Active Class Selection. In: Kok, J.N., Koronacki, J., Lopez de Mantaras, R., Matwin, S., Mladenič, D., Skowron, A. (eds.) ECML 2007. LNCS (LNAI), vol. 4701, pp. 640–647. Springer, Heidelberg (2007)
9. Maganti, H.K., Scherer, S., Palm, G.: A Novel Feature for Emotion Recognition in Voice Based Applications. In: Paiva, A.C.R., Prada, R., Picard, R.W. (eds.) ACII 2007. LNCS, vol. 4738, pp. 710–711. Springer, Heidelberg (2007)
10. Monteleoni.: Learning with Online Constraints: Shifting Concepts and Active Learning. PhD thesis, Massachusetts Institute of Technology (2006)
11. Rabiner, L.R.: Fundamentals of Speech Recognition. Prentice-Hall (1993)
12. Scherer, K.R., Johnstone, T., Klasmeyer, G.: Affective Science. In: Handbook of Affective Sciences - Vocal expression of emotion, ch. 23, pp. 433–456. Oxford University Press (2003)
13. Scherer, S.: Analyzing the User's State in HCI: From Crisp Emotions to Conversational Dispositions. PhD thesis. Ulm University (2011)
14. Settles.: Curious Machines: Active Learning with Structured Instances. PhD thesis, University of Wisconsin Madison (2008)
15. Settles, B.: Active learning literature survey. Computer Sciences Technical Report 1648. University of Wisconsin–Madison (2009)

16. Thiel, C., Scherer, S., Schwenker, F.: Fuzzy-Input Fuzzy-Output One-against-all Support Vector Machines. In: Apolloni, B., Howlett, R.J., Jain, L. (eds.) KES 2007, Part III. LNCS (LNAI), vol. 4694, pp. 156–165. Springer, Heidelberg (2007)
17. Tong.: Active Learning: Theory and Applications. PhD thesis. Stanford University (2001)
18. Wendt, B.: Analysen Emotionaler Prosodie, Hallesche Schriften zur Sprechwissenschaft und Phonetik, vol. 20. Peter Lang Internationaler Verlag der Wissenschaften (2007)
19. Wendt, B., Scheich, H.: The "Magdeburger Prosodie Korpus" - a spoken language corpus for fMRI-studies. In: Speech Prosody SProSIG 2002, pp. 699–701 (2002)
20. Zhu, X.: Semi-supervised learning literature survey. Technical Report 1530, Computer Sciences. University of Wisconsin-Madison (2005)

Semi-supervised Linear Discriminant Analysis Using Moment Constraints

Marco Loog[*,**]

Pattern Recognition Laboratory,
Delft University of Technology,
The Netherlands
m.loog@tudelft.nl,
prlab.tudelft.nl

Abstract. A semi-supervised version of Fisher's linear discriminant analysis is presented. As opposed to virtually all other approaches to semi-supervision, no assumptions on the data distribution are made, apart from the ones explicitly or implicitly present in standard supervised learning. Our approach exploits the fact that the parameters that are to be estimated in linear discriminant analysis fulfill particular relations that link label-dependent with label-independent quantities. In this way, the later type of parameters, which can be estimated based on unlabeled data, impose constraints on the former and lead to a reduction in variability of the label dependent estimates. As a result, the performance of our semi-supervised linear discriminant is expected to improve over that of its supervised equal and typically does not deteriorate with increasing numbers of unlabeled data.

1 Introduction

We devise a semi-supervised scheme tailored to classical, widely-used, maximum likelihood-based linear discriminant analysis (LDA) [14] (a.k.a. Fisher linear discriminant, or normal-based linear discriminant function [18]). The build on a principle that has been presented in an earlier work on semi-supervision for nearest mean classification [15]. It suggests to exploit known relationships between the class-specific parameters to be estimated in the learning phase and certain label-independent statistics. Enforcing these constraints during semi-supervision, yields label-dependent estimates that have smaller expected deviation from the true parameter value, which, in turn, lead to reduced classification errors. Where [15] presents a straightforward way to enforce labeled-unlabeled constraints merely involving class means and overall means, in this paper we show how to deal with known constraints on the second order moments. These moments are relevant to LDA and somewhat more difficult to deal with.

---

[*] Partly supported by the Innovational Research Incentives Scheme of the Netherlands Research Organization [NWO, VENI Grant 639.021.611].

[**] Secondary affiliation with the Image Group, University of Copenhagen, Denmark.

F. Schwenker and E. Trentin (Eds.): PSL 2011, LNAI 7081, pp. 32–41, 2012.

A key feature of the approach is that no assumptions beyond those intrinsically present in the parameters have to be made. This is in contrast with the major share of the current approaches to semi-supervised learning that stress the need for additional assumptions on the available data, such as the cluster assumption, the smoothness assumption, the assumption of low density separation, and the manifold assumption [6,20,24]. As soon as the underlying model assumptions do not fit the data, however, there is the real risk that adding unlabeled data leads to a severe deterioration of classification performance [7,8,19].

The next section continuous with an overview of closest related work and introduces, among others, so-called self-learning or self-training. Section 3 recapitulates [15], which discusses nearest mean classification and illustrates the suggested approach in some more detail. Subsequently, Section 4 provides a particular implementation of the main idea geared to semi-supervision for LDA. Section 5 provides experimental results on various real-world data sets in which our constrained approach is compared to regular LDA and self-learned LDA. Section 6 completes the paper, providing a discussion and conclusions.

2 Related Work

There are few works that focus on semi-supervised LDA in particular. The most relevant contributions come from statistics and have been published mainly in the sixties and seventies. Hartley and Rao [12] suggest to maximize the likelihood over all permutations of possible labelings for the unlabeled data. Realizing that this approach is infeasible, [16,17] proposes to follow an iteratively maximization of the likelihood. In a first step, the linear discriminant is trained on the labeled data only. This trained classifier is then used to label all unlabeled instances. Together with the already labeled data, all now-labeled data is employed to retrain the classifier based on which one can relabel the initially unlabeled data. This process of relabeling originally unlabeled data may be repeated until all these instances do not change label anymore.

The former is basically a form of what nowadays is also known as self-learning, self-training, Yarowsy's algortihm [22]. The appealing feature of self-learning is that it basically can be used to train any known classification scheme in a semi-supervised way. When dealing with density-based, generative classifiers, one can avoid the hard assignments in every iteration of self-learning, explicitly include a term for the unlabeled data in the probabilistic model, and maximize its likelihood. Typically, the use of an expectation maximization algorithm is necessary to optimize the parameters in these models [19]. The similarity between methods using self-learning and expectation maximization, in some cases equivalence even, has been noted in various papers, e.g. [1,3], and it is to no surprise that such approaches suffer from the same drawback: As soon as the underlying model assumptions do not fit the data, there is the risk that adding unlabeled data leads to a substantial decrease of classification performance [7,8,19].

An approach seemingly different from self-learning goes under the name of label propagation. It makes a smoothness assumption that is expressed in the supposition that closer data points tend to belong to the same class. Various instantiations of this idea exist, most of which are closely related to graph-based, manifold learning, or spectral clustering methods [21,23,5]. The propagation of label information through such graph structure can also be thought of as a particular instance of the iterative expectation maximization and self-learning methods, especially if the underlying classifier is, for instance, a Parzen or a k nearest neighbor classifier. Another, more explicit connection between self-learning and graph-based propagation methods can be found in [9].

We note that there are also semi-supervised approach to LDA as a dimensionality reduction technique. As we consider LDA as a classifier, we do not discuss these approach here except for the work in [10], which comes close to ours in some sense. It aims to improve the estimates of particular parameters by including unlabeled data in the estimation procedure as well. It notes that the Fisher criterion, which typically employs the between-class and within-class covariance can also be expressed in such a way that the total covariance matrix replaces one of the other two (cf. [11]). Obviously, the total covariance can be better estimated using all data, both labeled and unlabeled, which in turn might result in better performance of the dimensionality reduction scheme. Our work, however, aims at LDA for classification in which the total covariance does not directly play a role and therefore we cannot resort to the simple and straightforward suggestion made in [10]. Still, the possibility to have such an ameliorated estimate of a covariance matrix is also at the basis of our approach as will be clarified in Section 4. In the next section, however, we first provide a brief review of constrained nearest mean classification for semi-supervision [15].

3   Semi-supervised Nearest Means

The semi-supervised version of the nearest mean classifier (NMC) proposed in [15] is rather simple but seems effective nonetheless. Firstly, note that when employing an NMC, the K class means, m_i with $i \in \{1, \ldots, K\}$, and the overall mean of the data, m, fulfill the constraint

$$m = \sum_{i=1}^{K} p_i m_i \,, \tag{1}$$

where p_i is the prior of class i. Now, when we have additional unlabeled data, one can improve the estimate of m, because it does not depend on labeled data. In this case, the constraint in Eqation (1) is violated, however. The core idea in [15] is that one can get improved estimates of the class means by adapting them such that the constraint is fulfilled again.

The solution from [15] simply alters all K sample class means m_i by the same shift such that the overall sample mean $m' = \sum_{i=1}^{K} p_i m_i'$ of the shifted class means m_i' coincides with the total sample mean m_t. The latter overall mean has been obtained using all data, both labeled and unlabeled. More precisely, one performs the following update of the class means

$$m'_i = m_i - \sum_{i=1}^{K} p_i m_i + m_t \tag{2}$$

for which one can easily check that $\sum_{i=1}^{K} p_i m'_i$ indeed equals m_t.

4 Semi-supervised LDA

Equation (1) constrains the possible configurations that the class means can take on by linking label dependent parameters with label independent parameters. For LDA, an additional, known constraint that provides such parameter constraints equates the sum of the estimates of the between-class covariance matrix $\mathbf{B}$ and within-class covariance $\mathbf{W}$ to the total covariance over all data $\mathbf{T}$ (cf. [11]), i.e.,

$$\mathbf{T} = \mathbf{W} + \mathbf{B}, \tag{3}$$

where

$$\mathbf{T} := \frac{1}{N} \sum_{i=1}^{K} \sum_{j=1}^{N_i} (x_{i,j} - m)(x_{i,j} - m)^t = \frac{1}{N} \sum_{n=1}^{N} (x_n - m)(x_n - m)^t \tag{4}$$

in which $x_{i,j}$ is the jth feature vector from class i, N_i is the number of samples from class i, m is the estimated overall mean, and N is the total number of samples. The index n enumerates all instances in the data set and makes explicit that $\mathbf{T}$ can indeed be determined in a label-independent way. For the remaining variable in Equation (3), we have the following definitions.

$$\mathbf{W} := \sum_{i=1}^{K} p_i \mathbf{C}_i, \tag{5}$$

where $\mathbf{C}_i$ is the estimated covariance matrix for class i, and

$$\mathbf{B} := \sum_{i=1}^{K} p_i (m_i - m)(m_i - m)^t. \tag{6}$$

The parameters of interest for LDA are the class means m_i and the within-class covariance matrix $\mathbf{W}$ and we would like to estimate these parameters from both labeled and unlabeled data under the constraints provided by Equations (1) and (3). In both equations the left hand side is fixed and determined by all data available, both labeled and unlabeled. Let us denote the estimated total mean based on all of the data by μ and let the corresponding total covariance matrix be denoted by Θ; m and $\mathbf{T}$ are the corresponding mean and covariance based only on the labeled data. The matrices $\mathbf{W}$ and $\mathbf{B}$, the vectors m_i, and the scalars p_i are the free parameters to be determined.

Now, we suggest the following—in principle—very easy and effective solution. To start with, transform every labeled datum x as follows:

$$x \leftarrow \mathbf{\Theta}^{\frac{1}{2}} \mathbf{T}^{-\frac{1}{2}} (x - m) + \mu . \tag{7}$$

The transformation sees to it that the overall mean and covariance statistics of the labeled data match the respective statistics as measured on all data. That is, on the transformed data, the corresponding m and $\mathbf{T}$ equal μ and $\mathbf{\Theta}$, respectively. Finally, the crucial idea is to simply train a regular LDA in this transformed space. The estimates for m_i and $\mathbf{W}$ obtained in this way are our semi-supervised estimates.

By means of Equation (6) we can also determine the corresponding $\mathbf{B}$ in the transformed space and by construction we now have

$$\mu = \sum_{i=1}^{K} p_i m_i \tag{8}$$

and

$$\mathbf{\Theta} = \mathbf{W} + \mathbf{B} . \tag{9}$$

Because the transformation that is applied is affine, we can actually estimate the m_is and the $\mathbf{W}$ directly in the original space. Given the class means m_i' and the within-class covariance matrix $\mathbf{W}'$ determined on the labeled data only, we can write

$$m_i = \mathbf{\Theta}^{\frac{1}{2}} \mathbf{T}^{-\frac{1}{2}} (m_i' - m) + \mu \tag{10}$$

and

$$\mathbf{W} = \mathbf{\Theta}^{\frac{1}{2}} \mathbf{T}^{-\frac{1}{2}} \mathbf{W}' \mathbf{\Theta}^{\frac{1}{2}} \mathbf{T}^{-\frac{1}{2}} , \tag{11}$$

which expresses the m_i and $\mathbf{W}$ in terms of first and second order statistics in the original space.

Note that the foregoing transformation is not necessarily unique and it depends on the precise definition of what is meant by the square root of a matrix. One consequence of this is that the transformation might not have the proper invariance properties, e.g. it might not be invariant to linear transformations. For that reason, we always pre-whiten the data such that $\mathbf{\Theta} = \mathbf{I}$ or $\mathbf{T} = \mathbf{I}$ and determine the data transformation starting from the pre-whitened data.

Note also that in the foregoing, it has been tacitly assumed that the matrix $\mathbf{T}$ is not singular, which is rather restrictive considering that semi-supervision may especially be interesting in case there are only few labeled instances available. Therefore, in the remainder, we replace the inverse with the Moore-Penrose generalized matrix inverse even though more clever choices might be possible.

5   Experimental Setup and Results

We carried out several experiments to substantiate some of the earlier findings and claims and to potentially further our understanding of the novel semi-supervised approach. We are interested to what extent LDA can be improved by semi-supervision and a comparison is made to the standard, supervised setting and an LDA trained by means of self-learning [16,22]. Similar results are obtained by explicit EM approaches [12,19].

As it is not directly of interest to this work, we do not consider learning curves for the number of labeled observations. We experimented mainly with ten and a hundred labeled training objects in total. In all cases we made sure every class has at least one training sample. We do consider learning curves as a function of the number of unlabeled instances. This setting easily disclosed both the sensitivity of the self-learning to an abundance of unlabeled data and the improvements that may generally be obtained given various quantities of unlabeled data. The number of unlabeled objects considered in the main experiments are 2, 8, 32, 128, 512, 2048, and 8192.

In the experiments, we occasionally suffer from a singular within-class covariance matrix, whose inverse is needed to apply LDA. Like for $\mathbf{T}$, the problem is essentially solved by proper use of a Moore-Penrose generalized inverse for $\mathbf{W}$, making the appropriate subspace considerations. Another possibly would be to use a regularized version of $\mathbf{W}$ in the LDA.

Table 1. Basic properties of the nine real-world data sets

data set	number of objects	dimensionality	smallest class prior
haberman	306	3	0.26
ionosphere	351	33	0.36
parkinsons	195	22	0.25
pima	768	8	0.35
sonar	208	60	0.47
spect	267	22	0.21
spectf	267	44	0.21
transfusion	748	3	0.24
wdbc	569	30	0.37

Nine real-world data sets, all having two classes, are taken from the UCI Machine Learning Repository [2]. The UCI data sets used are haberman, ionosphere, parkinsons, pima, sonar, spect, spectf, transfusion, and wdbc for which some specifications can be found in Table 1.

On these, extensive experimentation has been implemented in which for every combination of number of unlabeled objects and labeled objects 1,000 repetitions were executed. In order to be able to do so on the limited amount of samples in the UCI data sets, we allowed to draw instances with replacement. This approach enabled us to properly study the influence of the constraint estimation on real-world data without having to deal with the extra variation due to cross validation or the like. This approach basically assumes that the empirical distribution of every data set is its true distributions and allows us to measure the error rates on the full data set. For different flexible classifiers this might give unacceptably

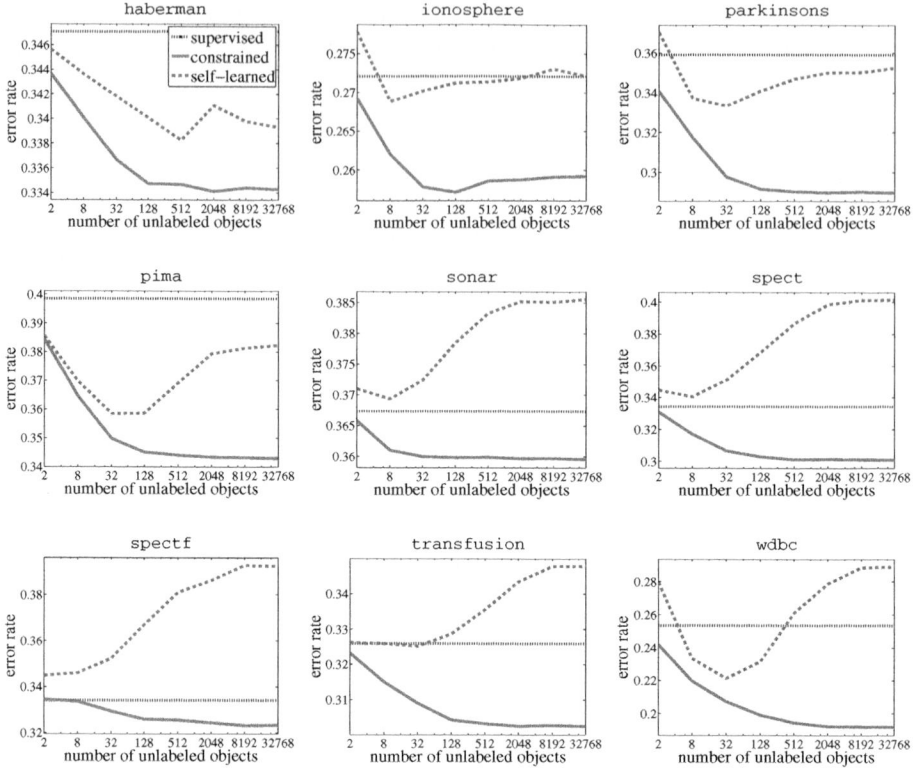

Fig. 1. Error rates for the supervised, semi-supervised, and self-learned classifiers on the nine real-world data sets for various unlabeled sample sizes and a total of ten labeled training samples

biased results. In our case, however, we are comparing different version of LDA, and we believe that the we can compare the various techniques based on these experiments.

Figures 1 and 2 provide the average learning curves for 10 and 20 labeled samples respectively. In the first place, one should notice that in most of the experiments the constrained LDA performs best of the three schemes employed. In addition, the self-learner leads to deteriorated performance with increasing unlabeled data sizes in close to all cases. An other interesting observation is that adding only a moderate amount of unlabeled objects already allows our semi-supervised constrained approach to outperform regular supervised LDA. In some cases our approach seems to perform worse, notably for `haberman` and `wdbc`. In those cases, however, the difference with supervised LDA is in the third or fourth decimal only.

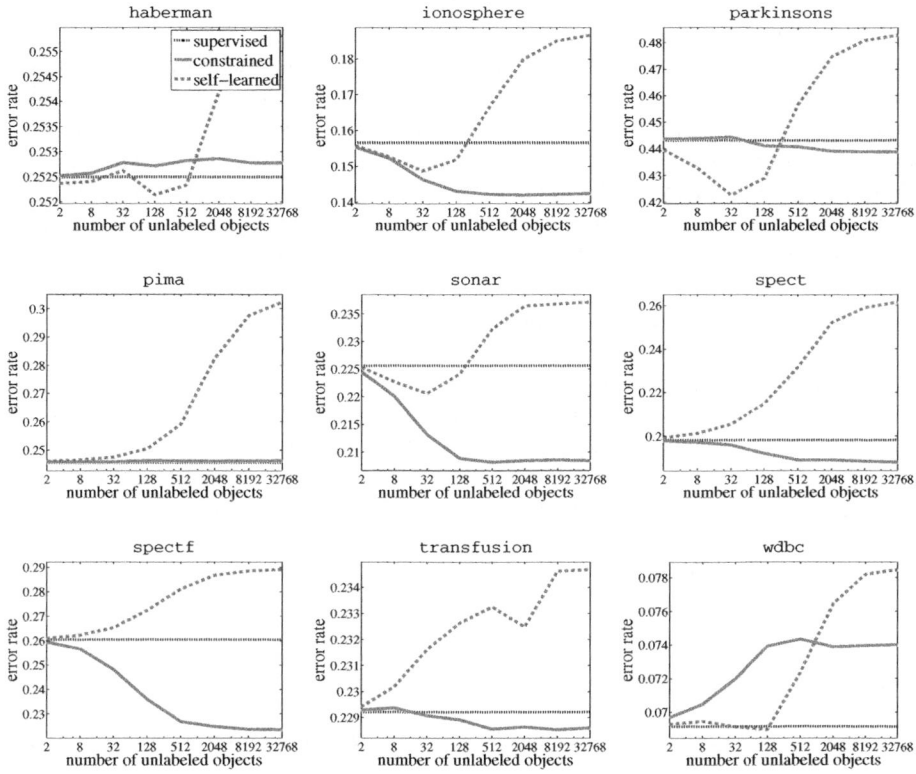

Fig. 2. Error rates for the supervised, semi-supervised, and self-learned classifiers on the nine real-world data sets for various unlabeled sample sizes and a total of a hundred labeled training samples

6 Discussion and Conclusion

We proposed to perform semi-supervised linear discriminant analysis (LDA) by making use of known constraints that link label-independent to label-dependent parameters (cf. [15]). Provided that the parameters that should be estimated in the learning phase of the classifier are part of these constraints together with the fact that the label-independent can be estimated more accurately when additional unlabeled data is available, this has the effect that also the parameters of interest can be estimated more precisely. A clear additional advantage is that our semi-supervised LDA is as easy to train as the regular LDA with no need for complex regularization schemes or iterative procedures as in [12,16,17,19]. We basically only need a preprocessing step that transforms the original data after which we can construct a standard LDA.

The proposed semi-supervised approach to LDA is a significant advance over the earlier proposed semi-supervised nearest mean classifier [15] as our approach allows to take into account important constraints on second-order moments and not only simple constraints on sample means.

The experiments show convincingly that rather substantial improvements can be obtained by enforcing the parameter constraints. This is contrasted with results obtained by self-learned LDA that often times performs dramatically worse, even compared to the performance of regular, supervised LDA. Nevertheless, we seem not to be there yet. Our current version of the semi-supervised constrained LDA may also lead to deteriorations in performance, however small they are. At this time, however, we do no clear idea about the reasons for these slight deteriorations.

The idea of linking label-dependent and label-independent estimates is of course more broadly applicable, the problem however is that it is not directly clear which constraints to apply to most of the other classical decision rules, if at all applicable. One of the main questions is if there is a general principle of constructing and applying such constraints. On the other hand, even if the principle would proof to be of limited applicability when dealing with other classifiers, one should realize that LDA acts as a basis for a broad class of penalized, flexible, and kernelized variations, as described in [4,13] and [14], for instance. Our procedure can be applied to these and as such provides a large class of semi-supervised classification schemes.

References

1. Abney, S.: Understanding the Yarowsky algorithm. Computational Linguistics 30(3), 365–395 (2004)
2. Asuncion, A., Newman, D.: UCI machine learning repository (2007), http://www.ics.uci.edu/~mlearn/MLRepository.html
3. Basu, S., Banerjee, A., Mooney, R.: Semi-supervised clustering by seeding. In: Proceedings of the Nineteenth International Conference on Machine Learning, pp. 19–26 (2002)
4. Baudat, G., Anouar, F.: Generalized discriminant analysis using a kernel approach. Neural Computation 12(10), 2385–2404 (2000)
5. Bengio, Y., Delalleau, O., Le Roux, N.: Label propagation and quadratic criterion. In: Semi-Supervised Learning, ch. 11. MIT Press (2006)
6. Chapelle, O., Schölkopf, B., Zien, A.: Introduction to semi-supervised learning. In: Semi-Supervised Learning, ch. 1. MIT Press (2006)
7. Cohen, I., Cozman, F., Sebe, N., Cirelo, M., Huang, T.: Semisupervised learning of classifiers: Theory, algorithms, and their application to human-computer interaction. IEEE Transactions on Pattern Analysis and Machine Intelligence, 1553–1567 (2004)
8. Cozman, F., Cohen, I.: Risks of semi-supervised learning. In: Semi-Supervised Learning, ch. 4. MIT Press (2006)
9. Culp, M., Michailidis, G.: An iterative algorithm for extending learners to a semi-supervised setting. Journal of Computational and Graphical Statistics 17(3), 545–571 (2008)
10. Fan, B., Lei, Z., Li, S.: Normalized LDA for semi-supervised learning. In: 8th IEEE International Conference on Automatic Face & Gesture Recognition, pp. 1–6. IEEE (2009)

11. Fukunaga, K.: Introduction to Statistical Pattern Recognition. Academic Press (1990)
12. Hartley, H., Rao, J.: Classification and estimation in analysis of variance problems. Review of the International Statistical Institute 36(2), 141–147 (1968)
13. Hastie, T., Buja, A., Tibshirani, R.: Penalized discriminant analysis. The Annals of Statistics 23(1), 73–102 (1995)
14. Hastie, T., Tibshirani, R., Friedman, J.: The Elements of Statistical Learning: Data Mining, Inference, and Prediction. Springer, Heidelberg (2001)
15. Loog, M.: Constrained Parameter Estimation for Semi-Supervised Learning: The Case of the Nearest Mean Classifier. In: Balcázar, J.L., Bonchi, F., Gionis, A., Sebag, M. (eds.) ECML PKDD 2010. LNCS (LNAI), vol. 6322, pp. 291–304. Springer, Heidelberg (2010)
16. McLachlan, G.: Iterative reclassification procedure for constructing an asymptotically optimal rule of allocation in discriminant analysis. Journal of the American Statistical Association 70(350), 365–369 (1975)
17. McLachlan, G.: Estimating the linear discriminant function from initial samples containing a small number of unclassified observations. Journal of the American Statistical Association 72(358), 403–406 (1977)
18. McLachlan, G.: Discriminant Analysis and Statistical Pattern Recognition. John Wiley & Sons (1992)
19. Nigam, K., McCallum, A., Thrun, S., Mitchell, T.: Learning to classify text from labeled and unlabeled documents. In: Proceedings of the Fifteenth National Conference on Artificial Intelligence, pp. 792–799 (1998)
20. Seeger, M.: A taxonomy for semi-supervised learning methods. In: Semi-Supervised Learning, ch. 2. MIT Press (2006)
21. Szummer, M., Jaakkola, T.: Partially labeled classification with Markov random walks. In: Advances in Neural Information Processing Systems, vol. 2, pp. 945–952 (2002)
22. Yarowsky, D.: Unsupervised word sense disambiguation rivaling supervised methods. In: Proceedings of the 33rd Annual Meeting on Association for Computational Linguistics, pp. 189–196 (1995)
23. Zhu, X., Ghahramani, Z.: Learning from labeled and unlabeled data with label propagation. Tech. Rep. CMU-CALD-02-107. Carnegie Mellon University (2002)
24. Zhu, X., Goldberg, A.: Introduction to Semi-Supervised Learning. Morgan & Claypool Publishers (2009)

Manifold-Regularized Minimax Probability Machine

Kazuki Yoshiyama and Akito Sakurai

School of Science for Open and Environmental System, Keio University
3-14-1 Hiyoshi, Kohoku-ku, Yokohama 223-8522, Japan
{k_yoshiyama,sakurai}@ae.keio.ac.jp

Abstract. In this paper we propose Manifold-Regularized Minimax Probability Machine, called MRMPM. We show that Minimax Probability Machine can properly be extended to semi-supervised version in the manifold regularization framework and that its kernelized version is obtained for non-linear case. Our experiments show that the proposed methods achieve results competitive to existing learning methods, such as Laplacian Support Vector Machine and Laplacian Regularized Least Square for publicly available datasets from UCI machine learning repository.

1   Introduction

The goal of semi-supervised learning is to utilize many unlabeled samples to improve generalization performance obtained from a few labeled samples. Many semi-supervised learning methods have been proposed recently from different viewpoints, such as density-based, cluster-based or graph-based (e.g. [3, 6, 16, 18]) and correspondingly by formulating different forms of loss function and/or regularization terms based on the original objective function.

In most of the proposals of these learning methods, they have extended existing learning methods to be used in semi-supervised setting. Seeing these approaches, we follow the way and extend minimax probability machine (MPM) to semi-supervised framework and propose a manifold-regularized minimax probability machine (MRMPM) based on graph-based regularization as explained in [3]. We also mention that the proposed MRMPM can be kernelized appropriately for non-linear case.

Our experiments show that our proposed methods achieve the competitive result to the existing manifold-regularized support vector machine and regularized least-square, called Laplacian Support Vector Machine (Lap-SVM) and Laplacian Regularized Least-Square (Lap-RLS) respectively, using the publicly available data sets from UCI machine learning repository.

The paper is organized as follows. In Section 2 we describe related works on which our proposed method is based. In Section 3 we describe a way to extend MPM to semi-supervised version and to derive our proposed method and show kernelized version can properly be obtained. In Section 4 we present empirical results of our experiments and compare the proposed method with the existing semi-supervised methods, Lap-SVM and Lap-RLS.

F. Schwenker and E. Trentin (Eds.): PSL 2011, LNAI 7081, pp. 42–51, 2012.

2   Related Work

In this section, we briefly summarize MPM proposed by [14] and explain Manifold Regularization framework proposed by [3] because our proposed method is significantly based on these. In Section 2.1 MPM is described and in Section 2.2 Manifold Regularization framework is explained and other graph-based works are mentioned.

2.1   Minimax Probability Machine

Consider the hyperplanes denoted by $\mathcal{H}(\mathbf{a}, b) = \{\mathbf{a}^T\mathbf{z} - b | \mathbf{a}, \mathbf{z} \in \mathbb{R}^d, b \in \mathbb{R}\}$, as in [14], which hopefully separate two classes $\mathcal{X}$ and $\mathcal{Y}$. MPM maximizes α, a lower bound of membership probability to each class with respect to all distributions having the prescribed means and covariance matrices. This is expressed as:

$$\max_{\alpha, \mathbf{a} \neq 0, b} \alpha \qquad s.t. \qquad \inf_{\mathbf{x} \sim (\overline{\mathbf{x}}, \Sigma_{\mathbf{x}})} \mathbf{Pr}\{\mathbf{a}^T\mathbf{x} \geq b\} \geq \alpha \qquad (1)$$

$$\inf_{\mathbf{y} \sim (\overline{\mathbf{y}}, \Sigma_{\mathbf{y}})} \mathbf{Pr}\{\mathbf{a}^T\mathbf{y} \leq b\} \geq \alpha,$$

where a random vector $\mathbf{x} \in \mathcal{X} \subset \mathbb{R}^d$, a mean vector $\overline{\mathbf{x}} \in \mathbb{R}^d$, and a positive definite covariance matrix $\Sigma_{\mathbf{x}} \in \mathbb{R}^{d \times d}$ in class $\mathcal{X}$; likewise for $\mathcal{Y}$. By exploiting Marshall and Olkin Theorem ([5, 14]), we can rewrite (1) as:

$$\max_{\alpha, \mathbf{a} \neq 0, b} \alpha \qquad s.t. \qquad b \leq \mathbf{a}^T\overline{\mathbf{x}} - \kappa(\alpha)\sqrt{\mathbf{a}^T\Sigma_{\mathbf{x}}\mathbf{a}} \qquad (2)$$

$$b \geq \mathbf{a}^T\overline{\mathbf{y}} + \kappa(\alpha)\sqrt{\mathbf{a}^T\Sigma_{\mathbf{y}}\mathbf{a}}.$$

where $\kappa(\alpha) = \sqrt{\frac{\alpha}{1-\alpha}}$. Since maximizing α is equivalent to maximizing $\kappa(\alpha)$, we can maximize κ without considering α. Further, considering the inequality constraints for b, we can eliminate b. Without loss of generality, we can set $\mathbf{a}^T(\overline{\mathbf{x}} - \overline{\mathbf{y}}) = 1$. Finally the problem (2) reduces to the optimization problem with respect to $\mathbf{a}$:

$$\kappa_*^{-1} = \min_{\mathbf{a}}(\|\Sigma_{\mathbf{x}}^{1/2}\mathbf{a}\|_2 + \|\Sigma_{\mathbf{y}}^{1/2}\mathbf{a}\|_2) \qquad s.t. \qquad \mathbf{a}^T(\overline{\mathbf{x}} - \overline{\mathbf{y}}) = 1 \qquad (3)$$

The problem (3) can be solved, i.e., we can obtain the optimal $\mathbf{a}_*$. Also, the optimal b can be computed as $b_* = \mathbf{a}_*^T\overline{\mathbf{x}} - \kappa\sqrt{\mathbf{a}_*^T\Sigma_{\mathbf{x}}\mathbf{a}_*}$

2.2   Manifold Regularization

For utilizing unlabeled samples, a manifold regularization framework was proposed by [3] which introduces a regularization that exploits the geometry of the marginal distribution. Suppose that there is a probability distribution P on $X \times \mathbb{R}$ according to which examples are generated. Labeled examples are (x, y) pairs generated according to P. Unlabeled examples are $x \in X$ generated according to the marginal distribution P_X of P. In the framework, the manifold regularization imposes a smoothness condition on possible solutions f by adding a penalty term $\|f\|_I^2$ where the norm is defined on a manifold $\mathcal{M}$, the support of P_X. Since P_X is unknown in most applications, an approximation based on the labeled and unlabeled data is considered. The choice in [3] is to use $\int_{\mathcal{M}} \langle \nabla_{\mathcal{M}} f, \nabla_{\mathcal{M}} f \rangle$ as $\|f\|_I^2$ and to approximate it on the basis of labeled and unlabeled data using the graph Laplacian. The graph here is an approximation to the manifold $\mathcal{M}$ where a node $\mathbf{x}$ in the graph is a point in $\mathcal{M}$ and the weight w_{ij} on an edge connecting two nodes $\mathbf{x}_i$ and $\mathbf{x}_j$ is the adjacency of the nodes. If $w_{ij} = \exp[-\sigma_s \|\mathbf{x}_i - \mathbf{x}_j\|_2^2]$, when the number of points goes infinity, after appropriate scaling, the graph Laplacian converges to the true Laplace-Beltrami operator on the manifold (Theorem 3.1 in [2]). Therefore we consider $\frac{1}{(\ell+u)^2} \sum_{i,j=1}^{\ell+u} w_{ij}(f(\mathbf{x}_i) - f(\mathbf{x}_j))^2$ in place of $\|f\|_I^2$.

Suppose that we are given a set of labeled samples $\{\mathbf{x}_i, y_i\}_{i=1}^{\ell}$ and a set of unlabeled samples $\{\mathbf{x}_j\}_{j=\ell+1}^{\ell+u}$ and $\|f\|^2$ is an appropriate smoothness condition on f in the function space of possible solutions, the optimization problem with the manifold regularization is:

$$\arg\min_f \frac{1}{\ell} \sum_{i=1}^{\ell} V(\mathbf{x}_i, y_i, f) + \gamma_A \|f\|^2 + \frac{\gamma_I}{(\ell+u)^2} \sum_{i,j=1}^{\ell+u} w_{ij}(f(\mathbf{x}_i) - f(\mathbf{x}_j))^2$$

$$= \arg\min_f \frac{1}{\ell} \sum_{i=1}^{\ell} V(\mathbf{x}_i, y_i, f) + \gamma_A \|f\|^2 + \frac{\gamma_I}{(\ell+u)^2} \mathbf{f}^T L \mathbf{f}, \tag{4}$$

where $\mathbf{f} = [f(\mathbf{x}_1), \ldots, f(\mathbf{x}_{\ell+u})]^T$, $V(\mathbf{x}_i, y_i, f)$ is some loss function, γ_A and γ_I control the complexity of f in the function space and in the intrinsic geometry of P_X respectively, and $L = D - W$ called graph Laplacian. Here W is the edge weights matrix of the data adjacency graph having the element w_{ij}, $D \in \mathbb{R}^{(\ell+u) \times (\ell+u)}$ is a diagonal matrix whose elements $\forall i$ $D_{ii} = \sum_{j=1}^{\ell+u} w_{ij}$, otherwise 0.

In this manifold regularization framework, by choosing squared loss $(y_i - f(\mathbf{x}_i))^2$ and hinge loss $\max[0, 1 - y_i f(\mathbf{x}_i)]$ as loss function for RLS and SVM respectively, RLS and SVM are extended to those semi-supervised versions, Lap-RLS and Lap-SVM [3].

Other related works based on the manifold-regularization framework are presented in [11, 17]. They incorporated dissimilarity into their objective function. In [7], the graph Laplacian was combined with Semi-Supervised Support Vector Machine. Also, smoothness measure analogous to the manifold regularization was used in [15].

3 Semi-supervised Minimax Probability Machine

In this section, we show that MPM can also be extended to Manifold-Regularized version and propose an algorithm, block coordinate descent, to solve the MRMPM optimization problem. Furthermore, we can obtain a kernelized version of MRMPM, called MRKMPM, based on a theorem similar to Corollary 5 in [14].

3.1   Manifold-Regularized Minimax Probability Machine

Our goal here is to construct MRMPM by utilizing the unlabeled samples. In order to incorporate the manifold regularization framework, we introduce a manifold-regularization term to the objective in the optimization problem (3) by following [3]. Therefore, the optimization problem (3) becomes the following optimization problem:

$$\kappa_*^{-1} = \min_{\mathbf{a}} \left(\|\Sigma_{\mathbf{x}}^{1/2}\mathbf{a}\|_2 + \|\Sigma_{\mathbf{y}}^{1/2}\mathbf{a}\|_2 + \frac{\gamma_I}{(\ell+u)^2} \sum_{i,j=1}^{\ell+u} w_{ij}(f(\mathbf{t}_i) - f(\mathbf{t}_j))^2 \right) \quad (5)$$
$$s.t. \quad \mathbf{a}^T(\overline{\mathbf{x}} - \overline{\mathbf{y}}) = 1,$$

where $\mathbf{t} \in \{\mathbf{x}_i\}_{i=1}^{N_x} \cup \{\mathbf{y}_i\}_{i=1}^{N_y} \cup \{\mathbf{z}_i\}_{i=1}^{N_z}$, where $\{\mathbf{z}\}_{i=1}^{N_z}$ are the unlabeled samples, and $f \in \mathcal{H}(\mathbf{a}, b)$. Since $(f(\mathbf{t}_i) - f(\mathbf{t}_j))$ in the problem (5) is equal to $(\mathbf{a}^T\mathbf{t}_i - \mathbf{a}^T\mathbf{t}_j)$, the optimization problem (5) can be rewritten to the following form:

$$\kappa_*^{-1} = \min_{\mathbf{a}}(\|\Sigma_{\mathbf{x}}^{1/2}\mathbf{a}\|_2 + \|\Sigma_{\mathbf{y}}^{1/2}\mathbf{a}\|_2 + \frac{\gamma_I}{(\ell+u)^2}\mathbf{a}^T ZLZ^T\mathbf{a}) \quad (6)$$
$$s.t. \quad \mathbf{a}^T(\overline{\mathbf{x}} - \overline{\mathbf{y}}) = 1,$$

where $Z \in \mathbb{R}^{d \times n}$ is a matrix composed of all labeled and unlabeled samples, $n = N_x + N_y + N_z$, and L is the graph Laplacian given by $L = D - W$.

Although the introduction of the manifold regularization term is straightforward, it is clear that the first and second terms appearing in the objective of (6) and the third term are different in scale and/or dimension. We therefore introduce the square root of the manifold-regularization term as our regularization term where the normalizing factor $\frac{1}{(\ell+u)^2}$ and regularization parameter γ_I are coerced to one parameter. Therefore, our optimization problem is:

$$\kappa_*^{-1} = \min_{\mathbf{a}}(\|\Sigma_{\mathbf{x}}^{1/2}\mathbf{a}\|_2 + \|\Sigma_{\mathbf{y}}^{1/2}\mathbf{a}\|_2 + \|\lambda^{1/2}M^{1/2}\mathbf{a}\|_2) \quad (7)$$
$$s.t. \quad \mathbf{a}^T(\overline{\mathbf{x}} - \overline{\mathbf{y}}) = 1,$$

where $M = ZLZ^T$ and λ is the regularization parameter. Since we add the manifold-regularization term to the problem (3), the expression of b_* in terms of $\mathbf{a}_*$ is changed. Before obtaining b_*, we trace back through the consideration on transforming (2) to (3), and obtain the counterpart of (2) which leads to the previous optimization problem (7):

$$\max_{\alpha, \mathbf{a} \neq 0, b} \alpha \quad s.t. \quad b \leq \mathbf{a}^T\overline{\mathbf{x}} - \kappa(\alpha)(\sqrt{\mathbf{a}^T\Sigma_{\mathbf{x}}\mathbf{a}} + \frac{1}{2}\sqrt{\lambda\mathbf{a}^T M\mathbf{a}}) \quad (8)$$
$$b \geq \mathbf{a}^T\overline{\mathbf{y}} + \kappa(\alpha)(\sqrt{\mathbf{a}^T\Sigma_{\mathbf{y}}\mathbf{a}} + \frac{1}{2}\sqrt{\lambda\mathbf{a}^T M\mathbf{a}}).$$

Comparing the optimization problem (8) with (2), one may think that the feasible region of α in the constraints of (8) is smaller than that of (2); however, we allow it since it is compensated by utilization of unlabeled samples. The optimization problem (8) indicates $b_* = \mathbf{a}_*^T\overline{\mathbf{x}} - \kappa_*(\sqrt{\mathbf{a}_*^T\Sigma_{\mathbf{x}}\mathbf{a}_*} + \frac{1}{2}\sqrt{\lambda\mathbf{a}_*^T M\mathbf{a}_*})$ by applying the same transformation of (2) to (3), to (8).

3.2 Algorithm

There is a transparent way to solve the optimization problem (3), i.e., block-coordinate descent ([10, 14]). We used the block-coordinate descent for the experiments described in the following. In what follows, we describe the block-coordinate descent for solving the MRMPM optimization problem.

The algorithm to solve MRMPM optimization problem is based on a simple iterative procedure, in which we make the optimization problem (7) unconstrained by writing $\mathbf{a} = \mathbf{a}_0 + F\mathbf{u}$, where $\mathbf{a}_0 = (\overline{\mathbf{x}} - \overline{\mathbf{y}})/\|\overline{\mathbf{x}} - \overline{\mathbf{y}}\|_2^2$, $\mathbf{u} \in \mathbb{R}^{d-1}$, and $F \in \mathbb{R}^{d \times (d-1)}$ whose columns is orthogonal to $(\overline{\mathbf{x}} - \overline{\mathbf{y}})$. Simultaneously, by squaring each term in the objective of (7) and introducing other variables to be optimized, we get the following form:

$$\inf_{\mathbf{u}, \beta, \eta, \gamma} \beta + \frac{1}{\beta}\|\Sigma_{\mathbf{x}}^{1/2}(\mathbf{a}_0 + F\mathbf{u})\|_2^2 + \eta + \frac{1}{\eta}\|\Sigma_{\mathbf{y}}^{1/2}(\mathbf{a}_0 + F\mathbf{u})\|_2^2 +$$
$$\gamma + \frac{\lambda}{\gamma}\|M^{1/2}(\mathbf{a}_0 + F\mathbf{u})\|_2^2. \tag{9}$$

The optimization problem (9) is equivalent to (7) because if we differentiate (9) with respect to β, η, and γ, we obtain $\beta_{k+1} = \sqrt{\mathbf{a}_k^T \Sigma_{\mathbf{x}} \mathbf{a}_k}, \eta_{k+1} = \sqrt{\mathbf{a}_k^T \Sigma_{\mathbf{y}} \mathbf{a}_k}$, and $\gamma_{k+1} = \sqrt{\lambda \mathbf{a}_k^T M \mathbf{a}_k}$ by setting each derivative as 0 at k-th iteration.

In the iterative procedure, we update β, η, γ as expressed above and $\mathbf{u}$ alternatively. Updating $\mathbf{u}$ is to solve least-square problem with respect to $\mathbf{u}$ by fixing β, η, and γ. Pseudo-code of the algorithm is presented in Algorithm 1.

Algorithm1	
Get Estimates	$\overline{\mathbf{x}}, \overline{\mathbf{y}}, \Sigma_{\mathbf{x}}, \Sigma_{\mathbf{y}}, M$
Compute	$\mathbf{a}_0 \leftarrow \mathbf{a}_0 + F\mathbf{u}$ F: columns of F are orthogonal to $(\overline{\mathbf{x}} - \overline{\mathbf{y}})$ $G \leftarrow F^T \Sigma_{\mathbf{x}} F$ $H \leftarrow F^T \Sigma_{\mathbf{y}} F$ $M \leftarrow \lambda F^T M F$ $\mathbf{g} \leftarrow F^T \Sigma_{\mathbf{x}} \mathbf{a}_0$ $\mathbf{h} \leftarrow F^T \Sigma_{\mathbf{x}} \mathbf{a}_0$ $\mathbf{m} \leftarrow \lambda F^T M \mathbf{a}_0$
Initialize	$\beta_1 = 1, \eta_1 = 1, \gamma_1 = 1, k = 1$
Repeat	$M_{LS} \leftarrow G/\beta_k + H/\eta_k + M/\gamma_k + \delta I$ $\mathbf{b}_{LS} \leftarrow -(\mathbf{g}/\beta_k + \mathbf{h}/\eta_k + \mathbf{m}/\gamma_k)$ solve $M_{LS}\mathbf{u}_k = \mathbf{b}_{LS}$ $w.r.t$ $\mathbf{u}_k$ $\mathbf{a}_k \leftarrow \mathbf{a}_0 + F\mathbf{u}_k$ $\beta_{k+1} \leftarrow \sqrt{\mathbf{a}_k^T \Sigma_{\mathbf{x}} \mathbf{a}_k}$ $\eta_{k+1} \leftarrow \sqrt{\mathbf{a}_k^T \Sigma_{\mathbf{y}} \mathbf{a}_k}$ $\gamma_{k+1} \leftarrow \sqrt{\lambda \mathbf{a}_k^T M \mathbf{a}_k}$ $k \leftarrow k + 1$
Until stop criterion satisfied or convergence	
Assign	$\mathbf{a} \leftarrow \mathbf{a}_k$ $\kappa \leftarrow 1/(\beta_k + \eta_k + \gamma_k)$ $b \leftarrow \mathbf{a}^T \overline{\mathbf{x}} - \kappa(\beta_k + \frac{1}{2}\gamma_k)$

3.3 Kernelization

We now proceed to kernelizing the MRMPM optimization problem (7). Following the conventional way where an inner product between nonlinear mappings is replaced by a given kernel, $k(\mathbf{t}_i, \mathbf{t}_j) = \phi(\mathbf{t}_i)^T \phi(\mathbf{t}_j)$ where $\phi : \mathbb{R}^d \mapsto \mathbb{R}^f$ and $\mathbb{R}^f$ is a higher-dimensional feature space, we first change $\mathbf{t}$ to $\phi(\mathbf{t})$, secondly obtain appropriate expressions of mean vectors and covariance matrices in order to make the inner product appear in a process of formulation, thirdly state a theorem similar to the representer theorem, and finally utilize the theorem for constructing the kernelized version of MRMPM.

First, let us consider in the mapped feature space a linear decision hyperplanes $\mathcal{H}(\mathbf{a}, b) = \{\mathbf{a}^T \phi(\mathbf{z}) - b | \mathbf{a}, \phi(\mathbf{z}) \in \mathbb{R}^f, b \in \mathbb{R}\}$ which separate two classes $\mathcal{X}$ and $\mathcal{Y}$ and the MRMPM problems described so far. Since the same discussion in the linear case is applicable to the nonlinear case, we obtain an optimization problem for the non-linear MRMPM

$$\kappa_*^{-1} = \min_{\mathbf{a}}(\|\Sigma_{\phi(\mathbf{x})}^{1/2}\mathbf{a}\|_2 + \|\Sigma_{\phi(\mathbf{y})}^{1/2}\mathbf{a}\|_2 + \|\lambda^{1/2}M_{\phi(\mathbf{t})}^{1/2}\mathbf{a}\|_2) \qquad (10)$$
$$s.t. \quad \mathbf{a}^T(\overline{\phi(\mathbf{x})} - \overline{\phi(\mathbf{y})}) = 1,$$

where $\overline{\phi(\mathbf{x})}, \overline{\phi(\mathbf{y})}, \Sigma_{\phi(\mathbf{x})}$, and $\Sigma_{\phi(\mathbf{y})}$ will be estimated in a similar way as explained in Theorem 1, $M_{\phi(\mathbf{t})} = \boldsymbol{\Phi}L\boldsymbol{\Phi}^T$, $\boldsymbol{\Phi} = (\phi(\mathbf{t}_1), \ldots, \phi(\mathbf{t}_n))$, L is graph Laplacian, and $\mathbf{t} \in \{\mathbf{x}_i\}_{i=1}^{N_x} \cup \{\mathbf{y}_i\}_{i=1}^{N_y} \cup \{\mathbf{z}_i\}_{i=1}^{N_z}$.

Next, let us state the following theorem parallel to Corollary 5 in [14].

Theorem 1. *Let $\{\mathbf{x}_i\}_{i=1}^{N_x}$, $\{\mathbf{y}_i\}_{i=1}^{N_y}$ be the labeled training samples in the classes corresponding to $\mathcal{X}$ and $\mathcal{Y}$ respectively and $\{\mathbf{z}_i\}_{i=1}^{N_z}$ be the unlabeled training samples. If $\overline{\mathbf{x}}, \overline{\mathbf{y}}, \Sigma_{\mathbf{x}}, \Sigma_{\mathbf{y}}$ can be written as*

$$\overline{\mathbf{x}} = \sum_{i=1}^{N_x} \lambda_i \mathbf{x}_i, \qquad \overline{\mathbf{y}} = \sum_{i=1}^{N_y} \omega_i \mathbf{y}_i,$$

$$\Sigma_{\mathbf{x}} = \rho_x I_d + \sum_{i=1}^{N_x} \Lambda_i(\mathbf{x}_i - \overline{\mathbf{x}})(\mathbf{x}_i - \overline{\mathbf{x}})^T,$$

$$\Sigma_{\mathbf{y}} = \rho_y I_d + \sum_{i=1}^{N_y} \Omega_i(\mathbf{y}_i - \overline{\mathbf{y}})(\mathbf{y}_i - \overline{\mathbf{y}})^T$$

where I_d is the identity matrix of dimension d, then the optimal $\mathbf{a}$ will lie in the span of the labeled samples $\{\mathbf{x}_i\}_{i=1}^{N_x}$, $\{\mathbf{y}_i\}_{i=1}^{N_y}$, and unlabeled samples $\{\mathbf{z}_i\}_{i=1}^{N_z}$.

Proof. We can write any $\mathbf{a}$ as $\mathbf{a}_d + \mathbf{a}_p$, where $\mathbf{a}_d$ is the projection of $\mathbf{a}$ in the span of the samples (vector space spanned by all samples $\{\mathbf{x}_i\}_{i=1}^{N_x}$, $\{\mathbf{y}_i\}_{i=1}^{N_y}$ and

$\{\mathbf{z}_i\}_{i=1}^{N_z}$), whereas $\mathbf{a}_p$ is the orthogonal component to the samples. One can then easily check that

$$\sqrt{\mathbf{a}^T \Sigma_{\mathbf{x}} \mathbf{a}} = \sqrt{\mathbf{a}_d \sum_{i=1}^{N_x} \Lambda_i (\mathbf{x}_i - \overline{\mathbf{x}})(\mathbf{x}_i - \overline{\mathbf{x}})^T \mathbf{a}_d + \rho_x (\mathbf{a}_d^T \mathbf{a}_d + \mathbf{a}_p^T \mathbf{a}_p)},$$

$$\sqrt{\mathbf{a}^T \Sigma_{\mathbf{y}} \mathbf{a}} = \sqrt{\mathbf{a}_d \sum_{i=1}^{N_y} \Omega_i (\mathbf{y}_i - \overline{\mathbf{y}})(\mathbf{y}_i - \overline{\mathbf{y}})^T \mathbf{a}_d + \rho_y (\mathbf{a}_d^T \mathbf{a}_d + \mathbf{a}_p^T \mathbf{a}_p)},$$

$$\sqrt{\mathbf{a}^T Z L Z^T \mathbf{a}} = \sqrt{\mathbf{a}_d^T Z L Z^T \mathbf{a}_d},$$

$$\mathbf{a}^T (\overline{\mathbf{x}} - \overline{\mathbf{y}}) = \mathbf{a}_d^T (\overline{\mathbf{x}} - \overline{\mathbf{y}}),$$

because $\mathbf{a}_p^T \mathbf{t}_i = 0$, $i \in \{1, 2, \ldots, N_x + N_y + N_z\}$ and $\mathbf{a}_d^T \mathbf{a}_p = 0$. Thus, an orthogonal component $\mathbf{a}_p$ of $\mathbf{a}$ won't affect the equality constraint in (7). Since the objective is to be minimized, we get $\mathbf{a}_p = 0$, therefore $\mathbf{a} = \mathbf{a}_d$: the optimal $\mathbf{a}$ will lie in the span of the labeled samples $\{\mathbf{x}_i\}_{i=1}^{N_x}$, $\{\mathbf{y}_i\}_{i=1}^{N_y}$, and unlabeled samples $\{\mathbf{z}_i\}_{i=1}^{N_z}$.   □

We can easily check that the optimization problem (7) can be entirely expressed in terms of the inner product between the samples only, so one can write $\mathbf{a}$ as a linear combination of the samples.

Finally, by applying Theorem 1 to samples mapped by non-linear mapping and using a simple algebraic manipulation, we can write the manifold-regularized kernelized version of MPM (MRKMPM) as follows:

$$\kappa_*^{-1} = \min_{\gamma} \quad \sqrt{\gamma^T L_{\mathbf{x}}^T L_{\mathbf{x}} \gamma} + \sqrt{\gamma^T L_{\mathbf{y}}^T L_{\mathbf{y}} \gamma} + \sqrt{\lambda \gamma^T K L K^T \gamma}$$

$$s.t. \quad \gamma^T (\mathbf{l}_{\mathbf{x}} - \mathbf{l}_{\mathbf{y}}) = 1 \tag{11}$$

where

$$L = \begin{pmatrix} K_{\mathbf{x}} - \mathbf{1}_{N_x} \mathbf{l}_{\mathbf{x}}^T \\ K_{\mathbf{y}} - \mathbf{1}_{N_y} \mathbf{l}_{\mathbf{y}}^T \end{pmatrix} = \begin{pmatrix} \sqrt{N_x} L_{\mathbf{x}} \\ \sqrt{N_y} L_{\mathbf{y}} \end{pmatrix}, K = \begin{pmatrix} K_{\mathbf{x}} \\ K_{\mathbf{y}} \\ K_{\mathbf{z}} \end{pmatrix}, K_{ij} = \phi(\mathbf{t}_i)^T \phi(\mathbf{t}_j)$$

$$(\mathbf{l}_{\mathbf{x}}^T)_i = \frac{1}{N_x} \sum_{j=1}^{N_x} K(\mathbf{x}_j, \mathbf{t}_i), (\mathbf{l}_{\mathbf{y}}^T)_i = \frac{1}{N_y} \sum_{j=1}^{N_y} K(\mathbf{y}_j, \mathbf{t}_i),$$

$$i \in \{1, 2, \ldots, N_x + N_y + N_z\}.$$

Note that K is arranged in order of the labeled samples $\{\mathbf{x}\}_{i=1}^{N_x}$, $\{\mathbf{y}\}_{i=1}^{N_y}$, and unlabeled samples $\{\mathbf{z}\}_{i=1}^{N_z}$, as $\mathbf{t} \in \{\mathbf{x}\}_{i=1}^{N_x} \cup \{\mathbf{y}\}_{i=1}^{N_y} \cup \{\mathbf{z}\}_{i=1}^{N_z}$, and $\mathbf{1}_m$ is a column vector having all ones and of dimension $m \in \{N_x, N_y\}$. The optimal b is obtained as $b_* = \gamma_*^T \mathbf{l}_{\mathbf{x}} - \kappa_* (\sqrt{\gamma_*^T L_{\mathbf{x}}^T L_{\mathbf{x}} \gamma_*} + \frac{1}{2} \sqrt{\lambda \gamma_*^T K L K^T \gamma_*})$. Once finding the optimal parameters γ, we can evaluate:

$$\text{sign}(a_*^T \phi(\mathbf{t}^{new}) - b_*) = \text{sign}(\sum_{i=1}^{N_x + N_y + N_z} [\gamma_*]_i K(\mathbf{t}_i, \mathbf{t}^{new}) - b_*).$$

If this value is $+1$, a new sample $\mathbf{t}$ is classified to class $\mathcal{X}$, otherwise to class $\mathcal{Y}$.

It is straightforward to solve the MRKMPM optimization problem (11) by modifying the Algorithm 1 for the non-linear case.

4 Experiment

In this section, we show the performances of our proposed MRMPM and MRKMPM in comparison with other semi-supervised learning methods, Lap-SVM, Lap-RLS, their kernelized versions Lap-KSVM and Lap-KRLS.

The datasets we used are ones from UCI machine learning repository. In our experiments we randomly selected 3 labeled samples in each class and 500 unlabeled samples (total number of training samples is 3 times of the number of classes plus 500), except for datasets which contain less than 500 samples. In such datasets we randomly selected 3 labeled samples in each class and used the rest samples as the unlabeled samples. These selected samples were used for training, the others and the unlabeled samples were used for test. The experiments in this setting were conducted 10 times for each dataset. For datasets which contain more than two classes, we use pairwise technique.

Regularization parameters γ_A, γ_I for Lap-SVM, Lap-RLS, Lap-KSVM, and Lap-KRLS, and λ for MRMPM and MRKMPM were set from $\{10^{-2}, 10^{-1}, 1, 10, 10^2\}$. We used RBF kernel as a kernel function in the from $k(\mathbf{x}_i, \mathbf{x}_j) = \exp\left[-\sigma_k \|\mathbf{x}_i - \mathbf{x}_j\|_2^2\right]$ and the same one in the same form as the edge weight in the adjacency graph. The hyperparameter σ_k in RBF kernel and the edge weight parameter σ_s in the graph Laplacian were set from $\{10^{-2}, 10^{-1}, 1, 10, 10^2\}$. We used the normalized graph Laplacian as $\tilde{L} = D^{-1/2}(D - W)D^{-1/2}$ instead of using graph Laplacian.

The best results in terms of average accuracies among the results for all possible combination of parameters described above in the experiments are reported with standard deviations in Table 1.

Table 1. The average accuracies with standard deviations. The accuracies are best in all the possible combinations of parameters tried in the experiments described in the main text. Three methods are compared in their linear and kernelized versions. The bold numbers mean best performance in the same row.

	MRMPM	MRKMPM	Lap-SVM	Lap-KSVM	Lap-RLS	Lap-KRLS
abalone	**52.37** (1.65)	48.36 (4.78)	49.60 (4.91)	48.84 (2.84)	52.15 (1.64)	49.94 (1.78)
glass	87.35 (14.37)	86.89 (15.42)	55.71 (12.63)	93.16 (2.57)	92.09 (3.81)	**96.02** (2.49)
liver	59.76 (10.59)	56.46 (8.81)	61.09 (5.87)	59.62 (4.77)	**62.63** (5.55)	59.56 (5.01)
magic	70.11(6.30)	**79.31**(10.57)	62.40(4.65)	64.85(4.47)	67.22(9.45)	64.84(4.61)
mammographic	**75.43**(16.63)	63.19(12.51)	67.14(14.27)	69.70(15.55)	69.65(14.73)	70.73(17.66)
spect	71.36 (12.43)	83.85 (32.05)	82.96 (1.81)	73.58 (16.51)	**84.63** (0.53)	74.85 (17.15)
spectf	70.11(19.69)	**79.31**(30.07)	33.91(23.53)	62.57(19.09)	59.69(17.28)	63.56(21.11)
usps	70.40(6.27)	65.37(12.87)	62.11(12.40)	**74.51**(7.47)	72.13(10.82)	74.31(7.37)
wine	82.60(16.05)	55.62(13.96)	63.96(13.70)	69.64(14.25)	**83.79**(14.14)	70.86(15.72)
simple average	71.05	68.71	59.88	68.50	71.55	69.41

Table 1 shows that our semi-supervised learning methods outperform the other semi-supervised learning methods in four datasets and achieve the competitive results in the other datasets. The table also shows that in linear versions, our method outperforms others in four datasets and is comparative in one dataset; in kernelized versions, ours wins three, Lap-KSVM wins two, and Lap-KRLS wins four out of nine datasets. If we take simple average of the accuracies for the nine datasets, we could see that our methods are comparative with Lap-RLS and Lap-KRLS and outperform Lap-SVM and Lap-KSVM. But clearly the standard deviations are so large that we could not derive statistically meaningful conclusions. The relatively high standard deviations in the results come from the fact that we only choose 3 labeled samples in each class.

5   Discussion

We proposed Manifold-Regularized Minimax Probability Machine and its kernelized version, for both of which theorems analogous to representer theorem hold. The experiments show that the proposed methods achieved competitive performances to the existing semi-supervised learning methods.

However, there remains several issues to be addressed. The manifold regularization framework helps to attain better generalization performance; however, when we estimate a mean vector and covariance matrix of a class we use only labeled samples, which may result in causing poorer generalization performance. In Algorithm 1 to solve MPM problems, we take inverse of a matrix, (M_{LS} in Algorithm 1). Usually matrix inversion takes $O(n^3)$ of time, where n is the number of columns (or rows) of a matrix, the complexity makes MRMPM inapplicable to large scale problems. In addition, we sometimes fail into the numerical instability problem when we take inverse of a matrix. Another issue is the number of parameters we have to tune. In our experiments we show the best performance for each dataset because the number of labeled samples we used was a few, where we could not employ usual parameter tuning techniques, e.g. cross-validation. In such situation the larger the number of parameters is, the harder we tune parameters.

Therefore, our future directions are three; 1) combining EM-algorithm and metric learning methods ([9, 12, 13]) with MPM to obtain better estimate of mean vectors and covariance matrices, which may lead to better generalization performance and small standard deviation in semi-supervised setting; 2) searching for other algorithms to solve MPM optimization problems more efficiently by approximating inverse of matrices; 3) exploring parameter tuning techniques which are not based on labeled samples only.

Acknowledgements. We would like to acknowledge financial support from Institutional Program for Young Researcher Overseas Visits.

References

[1] Argyriou, A., Micchelli, C.A., Pontil, M.: When is a representer theorem? vector versus matrix regularizers. Journal of Machine Learning Research 10, 2507–2529 (2009)

[2] Belkin, M., Niyogi, P.: Towards a Theoretical Foundation for Laplacian-Based Manifold Methods. In: Auer, P., Meir, R. (eds.) COLT 2005. LNCS (LNAI), vol. 3559, pp. 486–500. Springer, Heidelberg (2005)

[3] Belkin, M., Niyogi, P., Sindhwani, V.: Manifold regularization: A geometric framework for learning from labeled and unlabeled examples. Journal of Machine Learning Research 7, 2399–2434 (2006)

[4] Belkin, M., Niyogi, P., Sindhwani, V.: On manifold regularization. In: International Conference on Artificial Intelligence and Statistics, AISTATS (2005)

[5] Bertsimas, D., Popescu, I.: Optimal inequalities in probability theory: A convex optimization approach. SIAM Journal of Optimization 15(3), 780–804 (2005)

[6] Bousquet, O., Chapelle, O., Hein, M.: Measure based regularization. In: Advances in Neual Information Processing Systems, vol. 16 (2004)

[7] Chapell, O., Sindhwani, V., Keerthi, S.S.: Optimization techniques for semi-supervised support vector machine. Journal of Machine Learning Research 9, 203–233 (2008)

[8] Chapelle, O., Zien, A.: Semi-supervised classification by low density sepration. In: International Conference on Artificial Intelligence and Statistics, AISTATS (2005)

[9] Davis, J.V., Kulis, B., Jain, P., Dhillon, S.I.S.: Information-theoretic metric learning. In: Proceedings of the 24th International Conference on Machine Learning (2007)

[10] Bertsekas, D.P.: Nonlinear programming. Athena Scientific, Belmont (1999)

[11] Goldberg, A.B., Zhu, X., Wright, S.: Dissimilarity in graph-based semi-supervised classification. In: Eleventh International Conference on Artificial Intelligence and Statistics, AISTATS (2007)

[12] Jain, P., Kulis, B., Dhillon, I.: Inductive regularized learning of kernel functions. In: Advances in Neual Information Processing Systems (2010)

[13] Kulis, B., Sustik, M., Dhillon, I.: Learning low-rank kernel matrices. In: Proceedings, the 23rd International Conference on Machine Learning (2006)

[14] Lanckriet, G.R.G., Ghaoui, L.E., Bhattacharyya, C., Jordan, M.I.: A robust minimax approach to classification. Journal of Machine Learning Research 3, 555–582 (2002)

[15] Li, Z., Liu, J., Tang, X.: Pairwise constraint propagation by semidefinite programming for semi-supervised classification. In: Proceedings of the International Conference on Machine Learning (2008)

[16] Sindhwani, V., Niyogi, P., Belkin, M.: Beyond the point cloud: from transductive to semi-supervised learning. In: Proceedings of the 22nd International Conference on Machine Learning (2005)

[17] Tong, W., Jin, R.: Semi-supervised learning by mixed label propagation. In: Proceedings of the Twenty-Second AAAI Conference on Artificial Intelligence, AAAI (2007)

[18] Zhu, X., Ghahramani, Z., Lafferty, J.: Semi-supervised learning using gaussian fields and harmonic functions. In: Proceedings of the International Conference on Machine Learning (2003)

Supervised and Unsupervised Co-training of Adaptive Activation Functions in Neural Nets

Ilaria Castelli and Edmondo Trentin

Dipartimento di Ingegneria dell'Informazione
Università di Siena, via Roma 56, Siena, Italy
{castelli,trentin}@dii.unisi.it

Abstract. In spite of the nice theoretical properties of mixtures of logistic activation functions, standard feedforward neural network with limited resources and gradient-descent optimization of the connection weights may practically fail in several, difficult learning tasks. Such tasks would be better faced by relying on a more appropriate, problem-specific basis of activation functions. The paper introduces a connectionist model which features adaptive activation functions. Each hidden unit in the network is associated with a specific pair $(f(\cdot), p(\cdot))$, where $f(\cdot)$ (the very activation) is modeled via a specialized neural network, and $p(\cdot)$ is a probabilistic measure of the likelihood of the unit itself being relevant to the computation of the output over the current input. While $f(\cdot)$ is optimized in a supervised manner (through a novel backpropagation scheme of the target outputs which do not suffer from the traditional phenomenon of "vanishing gradient" that occurs in standard backpropagation), $p(\cdot)$ is realized via a statistical parametric model learned through unsupervised estimation. The overall machine is implicitly a co-trained coupled model, where the topology chosen for learning each $f(\cdot)$ may vary on a unit-by-unit basis, resulting in a highly non-standard neural architecture.

Keywords: Co-training, partially unsupervised learning, adaptive activation function.

1 Introduction

Neural networks are one of the most common models used in the machine learning community: they have been successfully used for regression, classification and function approximation tasks. In spite of their popularity and their nice theoretical properties, practical training difficulties are often met in severe learning tasks. Indeed, the model could require a high number of hidden units in order to perform well. This would lead to an architecture with a high number of free parameters, more difficult to train, prone to overfit the training data and to get stuck into poor local minima of the criterion function. In this paper we introduce a neural model having adaptive activation functions, learned during the training procedure itself. The aim is to define a more flexible model (yet possibly simpler overall) in which the hidden units can compute arbitrary functions. Learning

F. Schwenker and E. Trentin (Eds.): PSL 2011, LNAI 7081, pp. 52–61, 2012.

problems that would require a huge number of logistic activation functions are expected to turn up way simpler once their solution relies on a basis of "right", problem-specific activation functions. Since no such basis is known in advance, the approach we propose suggests learning the functions from scratch, according to the very nature of the data at hand. Each function is specialized over the input space by means of a well-defined likelihood criterion. This can be formalized by saying that each hidden unit in the model is associated with a pair $(f(\cdot), p(\cdot))$ where $f(\cdot)$ is the unit-specific, adaptive activation function, while $p(\cdot)$ is the corresponding likelihood measure.

Neural networks are usually trained over a supervised dataset defined as $\boldsymbol{D} = \left\{ \left(\boldsymbol{x}^k, \boldsymbol{y}^k \right), k = 1 \ldots N \right\}$, where $\boldsymbol{x}^k \in \mathbb{R}^d$ is a vector of observed features and $\boldsymbol{y}^k \in \mathbb{R}^n$ is a target vector. The net is taught to reproduce the target output $\boldsymbol{y}^k$ when the feature vector $\boldsymbol{x}^k$ is presented in input. Usually, a gradient descent algorithm, like backpropagation [1], is used in order to minimize the criterion function

$$C(\boldsymbol{w}) = \frac{1}{2} \sum_{k=1}^{N} \sum_{i=1}^{n} \left(y_i^k - \widetilde{y}_i^k(\boldsymbol{w}) \right)^2 \tag{1}$$

where $\boldsymbol{w}$ are the connection weights, y_i^k and $\widetilde{y}_i^k(\boldsymbol{w})$ are the target and the output of the i-th output unit of the network over k-th input pattern, respectively. Each unit j in layer L_l, receives an activation given by $a_j = \sum_{u \in L_{l-1}} q_u w_{ju}$, where q_u is the output of u-th unit in the previous layer and w_{ju} is the connection weight from unit $u \in L_{l-1}$ to unit $j \in L_l$. The output of the unit is computed applying an activation function $f_j(\cdot)$ to a_j, namely $o_j = f_j(a_j)$. According to the universal approximation theorem of neural networks [2], multilayer perceptrons (MLPs) having one hidden layer made of sigmoidal units (see figure 1a) are universal approximators on a compact subset of $\mathbb{R}^d$. Although this theorem guarantees the existence of a network able to approximate any function given at least one hidden layer and sigmoid transfer functions, the network may need an arbitrary amount of weights. From a practical point of view, the number of hidden units required could be arbitrarily high, leading to difficulties during the training phase and to limited generalization capability. In the following, we outline a viable way out relying on adaptive activation functions. It is worth noting that using such functions $f(\cdot)$ realized via connectionist models will not affect the overall network's capability of being a "universal approximator", due to theoretical results drawn from the investigation of non-sigmoid activation functions [3,4]. As we say, a probabilistic weighting strategy is used in order to train and apply $f(\cdot)$ within the overall learning machine. A unit-specific likelihood measure $p(\cdot)$ is associated with $f(\cdot)$, affecting its optimization and its contribution to the computation of the network outputs. A co-training procedure of a supervised model $f(\cdot)$ and a partially unsupervised model $p(\cdot)$ will emerge. To all practical ends, the underlying idea is that $p(\cdot)$ forces $f(\cdot)$ to focus on input patters that are likely to be drawn from a specific probability distribution (whilst standard backpropagation implicitly assumes a uniform distribution over all input patterns), simplifying the learning task by reducing it to easier sub-tasks whose support is homogeneous (meaning that it presents certain regularities).

The idea of learning activation functions while training the network has been investigated in [5] where Catmull-Rom splines are proposed. In so doing, a reduction in terms of model complexity is achieved, but the constraints imposed on the number of hidden units do not guarantee universal approximation. Conversely, if we do not impose any constraint on the number of hidden units and we use a MLP to model the activation functions, the whole model is still a universal approximator.

In order to realize a network of adaptive activation functions, a simple MLP architecture (the *outer* network) with a limited number m of hidden units is initialized and trained with backpropagation (BP) first. If the learning task is not trivial, the connectionist solution obtained this way is expected to be far suboptimal. Once BP training has been completed (e.g., when the generalization error evaluated over a validation set does not improve any longer) the hidden units are replaced with simple MLP architectures (the *inner* networks). The architecture of the inner networks may differ on a unit-by-unit basis. This results in a non-standard, not-fully-connected topology (figure 1b). Inner networks are then trained in order to contribute solving the overall learning problem. The algorithm for our trainable-adaptive multilayer perceptron (TA-MLP) is presented in detail in the next section. In the following we assume that the outer network has only one hidden layer, while the extension of the algorithm to deeper architectures is presented in the companion paper [6]. Furthermore, since the technique does not rely on straightforward BP of the partial derivatives of the error function w.r.t. the parameters, it does not suffer from the phenomenon of "vanishing gradient" which may be met in multilayer standard networks. It turns out that estimation of $p(\cdot)$ may take a variety of forms, either entirely unsupervised or partially-unsupervised. Explicit solution of the latter estimation problem is presented in the companion paper, where an experimental demonstration of the overall algorithm is given. Preliminary conclusions are drawn in Section 3.

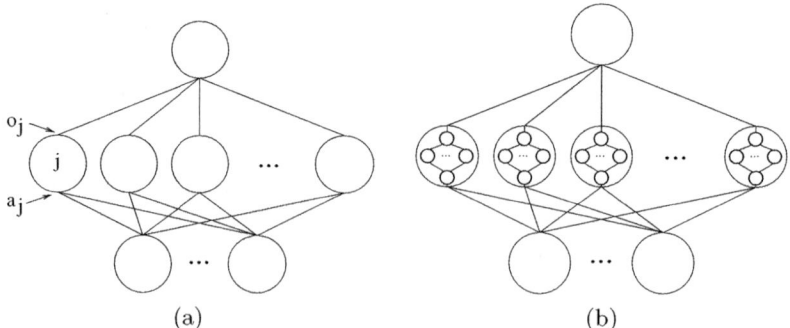

Fig. 1. (a) Classical MLP - (b) MLP with adaptive activation functions

2   The Training Algorithm

In order to train the inner networks, a training set $D_h = \left\{ \left(x_h^k, o_h^k \right), k = 1, \ldots, N \right\}$ must be specified for each net h, with $h = 1, \ldots, m$. Note that both x_h^k and o_h^k are scalar quantities. Regular backpropagation can then be applied. Section 2.1 elaborates on how D_h is generated. Once the creation of the network-specific training sets is accomplished, the probabilistic technique presented in Section 2.2 is applied for weighting individual input patterns on a network-by-network basis. Partially-supervised maximum-likelihood estimation of the quantities involved in the probabilistic weighting scheme is outlined in Section 2.3.

2.1   Generation of Locally-Supervised Training Sets

The k-th input pattern x_h^k for the h-th inner network can be easily obtained from its activation a_h:

$$x_h^k = \sum_{u \in L_0} x_u^k w_{hu} \qquad (2)$$

where x_u^k is the u-th entry of the original input vector x^k and L_0 is the input layer. More effort is required in order to define the target outputs. The supervision is available only at the output layer of the outer network, then it is necessary to define a strategy to back-propagate it. For each output unit i of the outer net, and for each pattern k, values of o_h^k are sought that satisfy the following equation:

$$\widetilde{y}_i^k = f_i \left(\sum_{h=1}^m o_h^k w_{ih} \right). \qquad (3)$$

First of all, we compute the target activations a_i of the output units of the outer network, by inversion of their activation function. In both cases of linear or sigmoidal activation function, computing the inverse is trivial. In the former case we have $y_i = f_i(a_i) = a_i$, that is $a_i = y_i$. If the activation is a sigmoid, i.e. $y_i = 1/(1 + \exp(-a_i))$, then $a_i = -\log(1 - y_i) + \log(y_i)$. In the latter case it is assumed that $y_i \in (0, 1)$. Then, the target activations a_i should be further backpropagated in order to compute the desired outputs o_h^k for each inner net. For clarity, the overall procedure is outlined in figure 2a and 2b. The former shows the trivial case where the outer network only has one output unit, while in the latter the straightforward extension to several output units is represented. Two different methods may be exploited: gradient descent and inversion of the weight matrix. Both are effective, and they generally lead to closely similar solutions. Of course, the target outputs o_h^k must be determined for each pattern in the training set, but for notational convenience we concentrate on a generic target o_h (i.e., from now on we drop the index k).

A gradient descent procedure can be exploited in almost exactly the same way as in the backpropagation algorithm. We are interested in minimizing the criterion function $C(\cdot)$ (see equation (1)) w.r.t. the output of the inner networks, o_h, with $h = 1, \ldots, m$. At this stage the weights between hidden and output

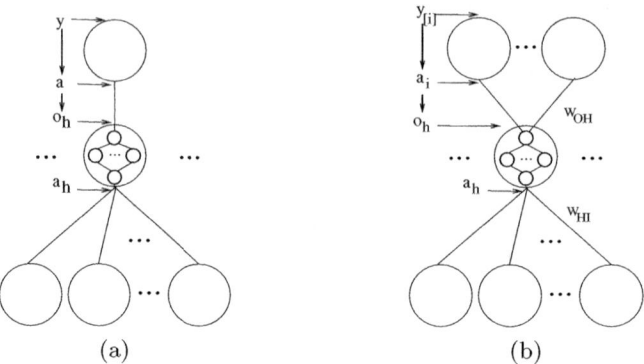

(a)   (b)

Fig. 2. Back-propagation of the target with (a) single a (b) multiple output units

layer of the outer MLP are considered to be constants (i.e., they are kept fixed at the values reached after the BP initialization of the outer network). At every iteration of BP the target output o_h of the generic h-th inner network is updated to a new value o'_h, according to the following rule:

$$o'_h = o_h + \Delta o_h = o_h - \eta \frac{\partial C}{\partial o_h} \qquad (4)$$

where η is the learning rate and, for instance, if the output activation is linear

$$
\begin{aligned}
\frac{\partial C}{\partial o_h} &= \frac{\partial}{\partial o_h} \left\{ \frac{1}{2} \sum_{i=1}^{n} (y_i - \widetilde{y}_i)^2 \right\} \\
&= \frac{1}{2} \sum_{i=1}^{n} \frac{\partial}{\partial o_h} (y_i - \widetilde{y}_i)^2 \\
&= \sum_{i=1}^{n} (y_i - \widetilde{y}_i) \frac{\partial}{\partial o_h} \left(y_i - \sum_{j=1}^{m} w_{ij} o_j \right) \\
&= - \sum_{i=1}^{n} (y_i - \widetilde{y}_i) w_{ih} \qquad (5)
\end{aligned}
$$

where n is the number of output units of the outer network, y_i and $\widetilde{y}_i$ are respectively the target and current output of the i-th output unit of the outer net.

A faster approach (albeit possibly less stable from a numeric standpoint) is provided by the inversion of the weight matrix of the outer network $W_{OH} \in \mathbb{R}^{n \times m}$ that connects the hidden to the output layer. Upon inversion of the output activation functions, we are provided with the array of desired activation at the n output units of the outer MLP, $A_O \in \mathbb{R}^n$. If $O_H \in \mathbb{R}^m$ is the desired array of outputs of the inner networks, then $A_O = W_{OH} O_H$ and

$$O_H = W_{OH}^{-1} A_O. \qquad (6)$$

Since the matrix W_{OH} usually does not have full rank, its pseudo-inverse is exploited. At this point, we have a generic, basic technique for creating the training sets for the inner networks. The next section investigates how a probabilistic weight $p(\cdot)$ is associated with each inner MLP. It will turn out that such probabilistic weights affect the very generation of target data, namely equation (5) and (6).

2.2 Probabilistic Weighting of Patterns

Since m hidden units are available, the original learning problem can be split into m smaller and easier tasks, and every inner net is specialized on one of such problems. This would be easily done having a method to evaluate the "competence" of each inner net on a given pattern. For this purpose, the posterior probability $P(h \mid x^k)$ of the h-th inner net given pattern x^k, can be exploited. Explicit calculation of $P(h \mid x^k)$ relies on a neuron-specific probability density function (pdf), namely $p_h(\cdot)$, that is the probabilistic quantity $p(\cdot)$ which we associate with each of the adaptive activation functions $f(\cdot)$ as anticipated in Section 1. In order to train inner networks we define a modified, neuron-specific criterion function in which every pattern x^k is weighted by $P(h \mid x^k)$, i.e. its probability of being in the region of competence of h-th inner net:

$$C_h\left(\boldsymbol{w_h}\right) = \frac{1}{2}\sum_{k=1}^{N} P(h \mid \boldsymbol{x^k})\left(o_h^k - \widetilde{o}_h^k\right)^2 \qquad (7)$$

where $\boldsymbol{w_h}$ are the connection weights of the inner net itself and $\widetilde{o}_h^k$ is its output. In so doing, the individual contribution each pattern x^k gives to the training of h-th inner MLP is proportional to the probability of the very MLP being competent over x^k. Probabilistic weighting are also exploited while computing target data for inner networks. Indeed, each inner net is expected to contribute to the activation of the output units of the outer net proportionally to $P(h \mid x^k)$. The weights from the hidden to the output units of the outer MLP can then incorporate the probabilistic weight. This means that, when the outer net is fed with pattern x^k, the activation of its i-th output unit is

$$a_i = \sum_{h=1}^{m} o_h^k w_{ih} P(h \mid \boldsymbol{x^k}) = \sum_{h=1}^{m} o_h^k \widetilde{w}_{ih} \qquad (8)$$

where we defined the variable weight $\widetilde{w}_{ih}$ (as a function of x^k) as $\widetilde{w}_{ih} = w_{ih}P(h \mid x^k)$. Computation of the target dataset for inner networks (see equation (6)) is then redefined in terms of these modified weights:

$$A_O = \widetilde{W}_{OH}O_H \quad \text{and} \quad O_H = \widetilde{W}_{OH}^{-1}A_O. \qquad (9)$$

Two examples drawn from a synthetic, one-dimensional regression task (generated as discussed in the companion paper [6]) are presented in figures 3a and 3b. The dots indicate the datasets obtained applying equation (9), while dashed lines represent the probabilistic weights themselves. Finally, note that the constraint imposed through equation (8) makes it possible to recover the target output during the feedforward phase. The next section outlines the steps for the actual calculation of $P(h \mid \boldsymbol{x}^k)$.

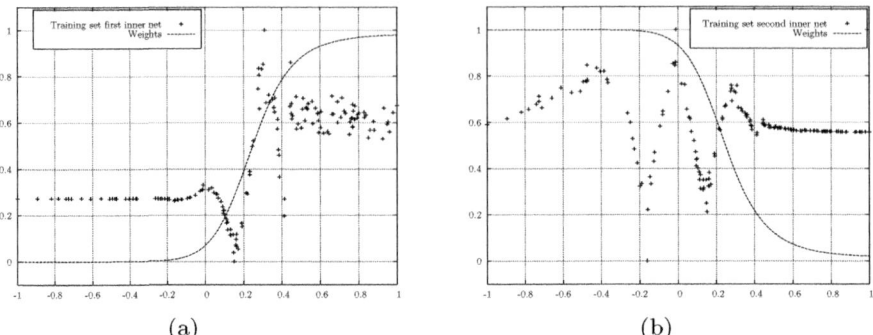

(a)   (b)

Fig. 3. Synthetic training set (dots) and the corresponding probabilistic weights (dashed line) generated for (a) first inner network and (b) second inner network

2.3 Partially Supervised Maximum-Likelihood Estimation of the Probabilistic Weights

In this section we point out how the probabilistic weights $P(h \mid \boldsymbol{x}^k)$ may be computed. Let us introduce a general, fully unsupervised framework first. Later on, we will extend the approach in a semi-supervised fashion, such that the probabilistic weights can be estimated by taking benefit from the knowledge of the neuron-specific target outputs during training. According to Bayes theorem the posterior probability $P(h \mid \boldsymbol{x})$ of the h-th inner net given the pattern $\boldsymbol{x}$ is

$$P(h \mid \boldsymbol{x}) = \frac{p(\boldsymbol{x} \mid h)P(h)}{p(\boldsymbol{x})}.$$

In practice, we associate a pdf $p_h(\boldsymbol{x}) = p(\boldsymbol{x} \mid h)$ with each adaptive neuron $h = 1, \ldots, m$. A classical Gaussian mixture model (GMM) can be used to estimate the likelihood term $p(\boldsymbol{x} \mid h)$ [1]. If we denote with $\boldsymbol{\theta}$ the parameters of the GMM, then

$$P(h \mid \boldsymbol{x}, \boldsymbol{\theta}) = \frac{p(\boldsymbol{x} \mid h, \boldsymbol{\theta})P(h)}{p(\boldsymbol{x} \mid \boldsymbol{\theta})}. \tag{10}$$

Assuming that the feature vectors $\boldsymbol{x}^1, \ldots, \boldsymbol{x}^N$ in the training sample are i.i.d. according to $p(\boldsymbol{x} \mid \boldsymbol{\theta})$, the likelihood of the parameters given the data is

$$p(\boldsymbol{x}^1, \ldots, \boldsymbol{x}^N \mid \boldsymbol{\theta}) = \prod_{k=1}^{N} p(\boldsymbol{x}^k \mid \boldsymbol{\theta})$$

where (following the usual GMM approach for a generic pattern $\boldsymbol{x}$)

$$p(\boldsymbol{x} \mid \boldsymbol{\theta}) = \sum_{j=1}^{c} P(\omega_j) p(\boldsymbol{x} \mid \omega_j, \boldsymbol{\theta_j})$$

$$= \sum_{j=1}^{c} P(\omega_j) \mathcal{N}(\boldsymbol{x}; \boldsymbol{\mu_j}, \Sigma_j) \qquad (11)$$

where c is the number of Gaussian components, $\boldsymbol{\mu_j}$, Σ_j and $P(\omega_j)$ are respectively the mean, the covariance matrix and the probability of the j-th component $\mathcal{N}(\boldsymbol{x}; \boldsymbol{\mu_j}, \Sigma_j)$, and $\boldsymbol{\theta_j} = (\boldsymbol{\mu_j}, \Sigma_j)$. Since we are interested in computing a probabilistic measure for each inner network h, we associate each activation function with a specific component density of the Gaussian mixture, i.e. $c = m$. In so doing, we are implicitly giving a rough probabilistic interpretation of the sigmoids realized by standard activation functions. In fact, the sigmoid (with a specific bias b and smoothness σ) is the cumulative distribution function of a corresponding logistic density function, that is close to a Gaussian distribution having mean b and variance $(\pi^2/3)\sigma^2$ (technically, the gap between multivariate Gaussian components and univariate distributions is going to be closed shortly). Standard maximum-likelihood estimation techniques can now be applied [6] in order to find $\boldsymbol{\theta_j}$, $j = 1, \ldots, m$, providing us with a complete algorithm.

So far, a viable and fully unsupervised approach has been outlined. A partially-supervised extension of the framework may benefit from the knowledge of the target outputs for the adaptive neurons during training. We perform an estimation of the joint probability of input and output data, i.e. instead of applying equation (11) we are interested in computing

$$p\left(\boldsymbol{x^k}, \boldsymbol{y^k} \mid \boldsymbol{\theta}\right) = \sum_{j=1}^{c} P(\omega_j) p(\boldsymbol{x^k}, \boldsymbol{y^k} \mid \omega_j, \boldsymbol{\theta_j}). \qquad (12)$$

A GMM can still be used. When the outer network is fed with pattern $\boldsymbol{x^k}$, the latter is projected first onto a subspace defined by the weight matrix W_{HI} (i.e., the connections between the input and the hidden layer). This defines the activations of the hidden units, that forms the input of the inner nets. Each Gaussian component is then defined on a different, univariate subspace, depending on the weights $\boldsymbol{w'_h} = (w_{h1}, w_{h2}, \ldots, w_{hd})$ connecting the input layer to the h-th adaptive hidden unit (see figure 4).

This translates in defining m univariate Gaussian probability density functions, each one defined on the subspace obtained applying to the input patterns the linear transformation given by the weights W_{HI}. Equation (12) is rewritten as

$$p\left(\boldsymbol{x^k}, \boldsymbol{y^k} \mid \boldsymbol{\theta}\right) = \sum_{j=1}^{c} P(\omega_j) p'(\boldsymbol{x^k}, \boldsymbol{y^k} \mid \omega_j, \boldsymbol{\theta_j}) \qquad (13)$$

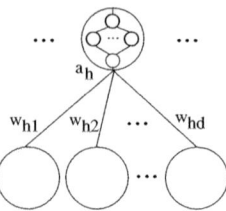

Fig. 4. Projection in hidden subspaces

where, referring to the h-th hidden unit,

$$p'(\boldsymbol{x}^k, \boldsymbol{y}^k \mid \omega_j, \boldsymbol{\theta}_j) = p(\boldsymbol{w}'_h \boldsymbol{x}^k, \boldsymbol{y}^k \mid \omega_j, \boldsymbol{\theta}_j). \qquad (14)$$

In this notation $\boldsymbol{w}'_h$ is meant to be a row vector and $\boldsymbol{x}^k$ is a column vector, then $\boldsymbol{w}'_h \boldsymbol{x}^k$ is a scalar quantity. If we let $\boldsymbol{z}^k = \left(\boldsymbol{w}'_h \boldsymbol{x}^k, \boldsymbol{y}^k\right)$ then we can rewrite equation (13) in the form

$$p\left(\boldsymbol{x}^k, \boldsymbol{y}^k \mid \boldsymbol{\theta}\right) = \sum_{j=1}^{c} P(\omega_j) p(\boldsymbol{z}^k \mid \omega_j, \boldsymbol{\theta}_j) \qquad (15)$$

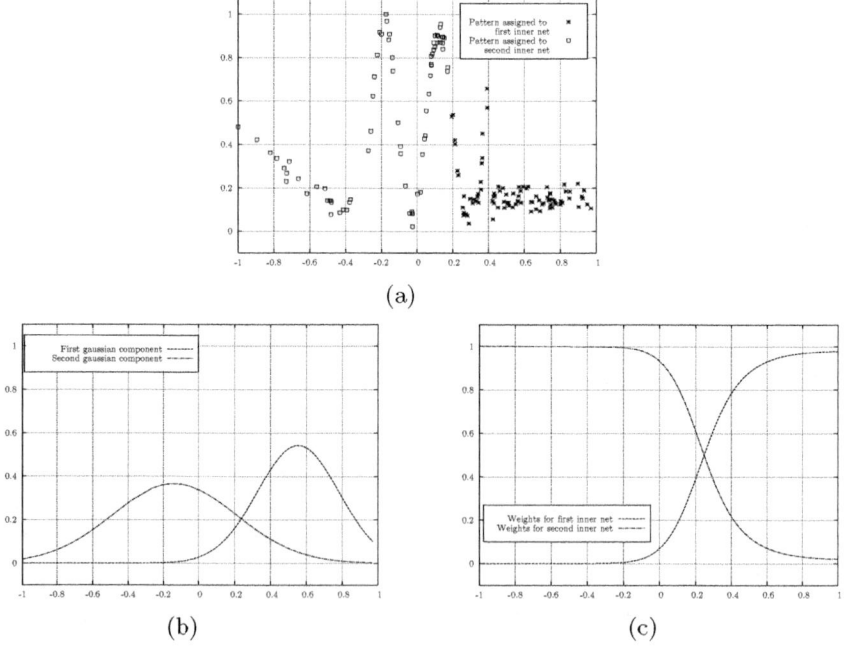

Fig. 5. (a) Pattern assigned to each Gaussian component after EM - (b) Gaussian components projected in input space - (c) probabilistic weights

An explicit maximum-likelihood solution of this estimation problem (including the probabilistic weighting of pattern we outlined in Section 2.2) based on the expectation-maximization (EM) algorithm, is developed in the companion paper [6]. During the test phase, the target y^k is not available, and then it is not possible to compute the exact value of $P(h \mid x^k, y^k)$. In practice, we project the Gaussian components in the original input space.

A graphic, illustrative example (taken from the same regression task plotted in figure 3a and 3b) is given in figure 5. Figure 5a shows the partition of the input patterns after running the EM algorithm. Each pattern x^k is assigned to the Gaussian component h for which $P(h \mid x^k)$ is higher. Figures 5b and 5c show respectively the two Gaussian components projected in input space and the probabilistic weights (the posterior probabilities $P(h \mid x)$ for $h = 1, 2$).

3   Preliminary Conclusions

The paper introduced the idea of adaptive activation functions in order to improve the learning capability of ANNs. A general form for the gradient-based training algorithm was outlined. Each adaptive activation is associated with a probabilistic measure $p(\cdot)$. Estimation of the latter may take place according to a standard, unsupervised maximum-likelihood, or in a partially unsupervised framework which exploits the joint pdf of feature vectors and target outputs. Explicit solution of the ML estimation in the latter scenario are developed in the companion paper [6], where the extension of the algorithm to multi-layer architectures is pointed out, too, and an experimental demonstration of the proposed model is given.

References

1. Duda, R.O., Hart, P.E., Stork, D.G.: Pattern Classification. Wiley (2001)
2. Cybenko, G.: Approximation by Superpositions of a Sigmoidal Function. Mathematics of Control, Signals, and Systems 4, 303–314 (1989)
3. Stinchcombe, M., White, H.: Universal Approximation using Feedforward Networks with Non-Sigmoid Hidden Layer Activation Functions. In: International Joint Conference on Neural Networks, IJCNN 1989, vol. 1, pp. 613–617 (1989)
4. Chen, T., Chen, H.: Universal Approximation to Nonlinear Operators by Neural Networks with Arbitrary Activation Functions and its Application to Dynamical Systems. IEEE Transaction on Neural Networks 4, 911–917 (1995)
5. Vecci, L., Piazza, F., Uncini, A.: Learning and Approximation Capabilities of Adaptive Spline Activation Function Neural Networks (1998)
6. Castelli, I., Trentin, E.: Semi-unsupervised Weighted Maximum-Likelihood Estimation of Joint Densities for the Co-Training of Adaptive Activation Functions. In: Schwenker, F., Trentin, E. (eds.) PSL 2011. LNCS (LNAI), vol. 7081, pp. 62–71. Springer, Heidelberg (2012)

Semi-unsupervised Weighted Maximum-Likelihood Estimation of Joint Densities for the Co-training of Adaptive Activation Functions

Ilaria Castelli and Edmondo Trentin

Dipartimento di Ingegneria dell'Informazione
Università di Siena, via Roma 56, Siena, Italy
{castelli,trentin}@dii.unisi.it

Abstract. The paper presents an explicit maximum-likelihood algorithm for the estimation of the probabilistic-weighting density functions that are associated with individual adaptive activation functions in neural networks. A partially unsupervised technique is devised which takes into account the joint distribution of input features and target outputs. Combined with the training algorithm introduced in the companion paper [2], the solution proposed herein realizes a well-defined, specific instance of the novel learning machine. The extension of the overall training method to more-than-one hidden layer architectures is pointed out, as well. A preliminary experimental demonstration is given, outlining how the algorithm works.

Keywords: Expectation maximization, partially unsupervised learning, co-training, adaptive activation function.

1 Introduction

In the companion paper an extension of the multilayer perceptron (MLP), named trainable-activations multilayer perceptron (TA-MLP), is introduced [2]. A TA-MLP is a flexible neural model having adaptive activation functions learned during the training procedure. The hidden units can compute task-specific arbitrary functions, learned according to the nature of the data. Each of them specializes over the input space according to a probabilistic criterion. The latter can be formalized by associating a pair $(f(\cdot), p(\cdot))$ with each hidden unit in the model, where $f(\cdot)$ is the adaptive activation function and $p(\cdot)$ is the corresponding likelihood measure. The quantity $f(\cdot)$ is realized by means of a MLP.

A partially-unsupervised probabilistic framework is used in order to let each hidden unit specialize on a part of the original problem. Each hidden unit h contributes to the output according to the probability $P(h \mid \boldsymbol{x})$ of that unit being "competent" on pattern $\boldsymbol{x}$. As explained in [2], a maximum-likelihood estimation of the parameters of a Gaussian mixture model (GMM) is required in order to compute $P(h \mid \boldsymbol{x})$. The GMM is expected to have as many component

F. Schwenker and E. Trentin (Eds.): PSL 2011, LNAI 7081, pp. 62–71, 2012.

densities as the number of hidden units in the TA-MLP (basically, each neuron specializes over a Gaussian distribution). The estimate of the GMM is then used within Bayes theorem in order to determine $P(h \mid x)$ [2].

As we say, each adaptive activation function relies on a standard MLP, called inner network. Flexibility of the overall learning machine can be increased further (e.g., when facing severe learning tasks) by replacing the inner net, in turn, with a TA-MLP. This may be applied recursively, as many times as necessary. In so doing, multiple levels of model expansion and adaptation are obtained. At each level, estimation of a new GMM is needed. Given an inner network h, let us call g_h its g-th hidden unit. When estimating $P(g_h \mid x)$, i.e. the posterior probability of the g_h-th inner network (within the h-th hidden unit of the outer MLP) given its input x, we need to take into account the probabilistic weight introduced at the previous level(s), i.e. $P(h \mid x)$. Then, maximum likelihood estimation (after the very first level) involves a weighting factor inherited from the previous levels. In Section 2 we introduce a simple refinement of the usual expectation-maximization (EM) algorithm for the estimation of GMM parameters [1] that accounts for this peculiar "pattern weighting" mechanism. Furthermore, as observed in [2], calculations occur in the joint input-output space at training time (taking benefit from knowledge of the target outputs). On the other way around, at test time the optimal parameters are projected back onto the bare input subspace. The overall training algorithm emerging from the combination of the general scheme proposed in [2] and the weighted estimation technique introduced below can be further extended to more-than-one hidden later (outer) MLPs, as well. The complete algorithm is handed out in Section 3. A preliminary experimental demonstration of how the TA-MLP works is given in Section 4, while Section 5 draws some conclusive remarks.

2 Maximum Likelihood Estimation with Weighted Patterns

Let us define a dataset $D = \left\{ \left(x^k, y^k \right), k = 1, \ldots, N \right\}$, where $x^k \in \mathbb{R}^d$ is a vector of observed features and $y^k \in \mathbb{R}^n$ is a target vector. In our partially-supervised framework we take benefit from the knowledge of the target outputs during training [2], and we define $z^k = \left(w_h' x^k, y^k \right)$, where $w_h' = (w_{h1}, w_{h2}, \ldots, w_{hd})$ is the vector of weights that connect the input layer to the h-th adaptive hidden unit. In this notation w_h' is meant to be a row vector and x^k is a column vector, that is, $w_h' x^k$ is a scalar quantity. Then, let us define the dataset $D' = \left\{ z^k, k = 1, \ldots, N \right\}$. In what follows we will work with the generic h-th inner net, i.e. all the input patterns x^k, $k = 1, \ldots, N$, are projected onto the subspace defined by the weights w_h'. Assuming that $z^1, \ldots, z^N$ are i.i.d., the likelihood of the parameters given D' is

$$p(D' \mid \theta) = \prod_{k=1}^{N} p(z^k \mid \theta) \qquad (1)$$

and $p(z^k \mid \boldsymbol{\theta})$ is expressed as a Gaussian mixture model (GMM):

$$p(z^k \mid \boldsymbol{\theta}) = \sum_{j=1}^{c} P(\omega_j) p(z^k \mid \omega_j, \boldsymbol{\theta}_j). \tag{2}$$

where $\boldsymbol{\theta}_j = (\boldsymbol{\mu}_j, \Sigma_j)$ are the parameters of the j-th Gaussian component (i.e. the mean and the covariance matrix) and $P(\omega_j)$ is its mixing parameter. If each pattern z^k has a weight v^k associated to it (in our case it is the probabilistic weight), equation (2) becomes

$$p(z^k \mid \boldsymbol{\theta}) = v^k \sum_{j=1}^{c} P(\omega_j) p(z^k \mid \omega_j, \boldsymbol{\theta}_j). \tag{3}$$

We can write the log-likelihood as

$$\log p(\boldsymbol{D}' \mid \boldsymbol{\theta}) = \sum_{k=1}^{N} \log \left\{ p(z^k \mid \boldsymbol{\theta}) \right\}. \tag{4}$$

In order to optimize the log-likelihood w.r.t. its parameters $\boldsymbol{\theta}$ the estimation of the optimal parameters for each component of the mixture is needed. We assume that $\boldsymbol{\theta}_i$ is functionally independent from $\boldsymbol{\theta}_j$ when $i \neq j$. We assume also identifiability of the components of the mixture, i.e. $\boldsymbol{\theta} \neq \tilde{\boldsymbol{\theta}} \Rightarrow \exists z \in \boldsymbol{D}'$: $p(z \mid \boldsymbol{\theta}) \neq p(z \mid \tilde{\boldsymbol{\theta}})$. Then, we compute the gradient of equation (4) w.r.t. the parameters of the generic i-th component, $\boldsymbol{\theta}_i$, and set it equal to zero:

$$\begin{aligned}
\nabla_{\boldsymbol{\theta}_i} \log p(\boldsymbol{D}' \mid \boldsymbol{\theta}) &= \sum_{k=1}^{N} \frac{1}{p(z^k \mid \boldsymbol{\theta})} \nabla_{\boldsymbol{\theta}_i} \left\{ v^k \sum_{j=1}^{c} P(\omega_j) p(z^k \mid \omega_j, \boldsymbol{\theta}_j) \right\} \\
&= \sum_{k=1}^{N} \frac{1}{p(z^k \mid \boldsymbol{\theta})} \nabla_{\boldsymbol{\theta}_i} \left\{ v^k P(\omega_i) p(z^k \mid \omega_i, \boldsymbol{\theta}_i) \right\} \\
&= \sum_{k=1}^{N} \frac{v^k P(\omega_i)}{p(z^k \mid \boldsymbol{\theta})} \nabla_{\boldsymbol{\theta}_i} \left\{ p(z^k \mid \omega_i, \boldsymbol{\theta}_i) \right\} \\
&= \sum_{k=1}^{N} \frac{v^k P(\omega_i \mid z^k, \boldsymbol{\theta})}{p(z^k \mid \omega_i, \boldsymbol{\theta}_i)} \nabla_{\boldsymbol{\theta}_i} \left\{ p(z^k \mid \omega_i, \boldsymbol{\theta}_i) \right\} \\
&= \sum_{k=1}^{N} v^k P(\omega_i \mid z^k, \boldsymbol{\theta}) \nabla_{\boldsymbol{\theta}_i} \log \left\{ p(z^k \mid \omega_i, \boldsymbol{\theta}_i) \right\} = \mathbf{0} \tag{5}
\end{aligned}$$

where $\mathbf{0}$ is a vector whose entries are all equal to zero. Compared to the classical unweighted estimation, we have the additional weighting factors v^k. Since each component of the mixture is Gaussian, $\boldsymbol{\theta}_j = (\boldsymbol{\mu}_j, \Sigma_j)$, we have:

$$\log p(z^k \mid \omega_j, \boldsymbol{\theta}_j) = -\log \left\{ (2\pi)^{d/2} \mid \Sigma_j \mid^{1/2} \right\} - \frac{1}{2} (z^k - \boldsymbol{\mu}_j)^\mathsf{T} \Sigma_j^{-1} (z^k - \boldsymbol{\mu}_j). \tag{6}$$

Taking the gradient of (6) w.r.t. $\boldsymbol{\mu_j}$ we obtain

$$\nabla_{\boldsymbol{\mu_j}} \log p(\boldsymbol{z^k} \mid \omega_j, \boldsymbol{\theta_j}) = \Sigma_j^{-1}(\boldsymbol{z^k} - \boldsymbol{\mu_j}) \tag{7}$$

and equation (5) can be rewritten as

$$\sum_{k=1}^{N} P(\omega_j \mid \boldsymbol{z^k}, \boldsymbol{\theta}) v^k \Sigma_j^{-1}(\boldsymbol{z^k} - \boldsymbol{\mu_j}) = \mathbf{0} \tag{8}$$

that is a set of $d+d^2$ equations that represent necessary conditions to be satisfied by the maximum-likelihood estimator. It follows that

$$\sum_{k=1}^{N} P(\omega_j \mid \boldsymbol{z^k}, \boldsymbol{\theta}) v^k \Sigma_j^{-1} \boldsymbol{z^k} = \sum_{k=1}^{N} P(\omega_j \mid \boldsymbol{z^k}, \boldsymbol{\theta}) v^k \Sigma_j^{-1} \boldsymbol{\mu_j} \tag{9}$$

and then

$$\boldsymbol{\mu_j} = \frac{\sum_{k=1}^{N} P(\omega_j \mid \boldsymbol{z^k}, \boldsymbol{\theta}) v^k \boldsymbol{z^k}}{\sum_{k=1}^{N} P(\omega_j \mid \boldsymbol{z^k}, \boldsymbol{\theta}) v^k} \tag{10}$$

In a similar manner, the gradient of (6) w.r.t. Σ_j can be calculated, yielding:

$$\Sigma_j = \frac{\sum_{k=1}^{N} P(\omega_j \mid \boldsymbol{z^k}, \boldsymbol{\theta}) v^k (\boldsymbol{z^k} - \boldsymbol{\mu_j})(\boldsymbol{z^k} - \boldsymbol{\mu_j})^{\mathsf{T}}}{\sum_{k=1}^{N} P(\omega_j \mid \boldsymbol{z^k}, \boldsymbol{\theta}) v^k}. \tag{11}$$

Finally, the mixing coefficients can be calculated taking the gradient of the log-likelihood w.r.t. $P(w_j)$ while imposing the constraint $\sum_{i=1}^{c} P(w_i) = 1$. This can be done using a Lagrange multiplier and maximizing the quantity

$$L = \sum_{k=1}^{N} \log \left\{ p(\boldsymbol{z^k} \mid \boldsymbol{\theta}) \right\} + \lambda \left(\sum_{i=1}^{c} P(\omega_i) - 1 \right)$$

$$= \sum_{k=1}^{N} \log \left\{ v^k \sum_{i=1}^{c} P(\omega_i) p(\boldsymbol{z^k} \mid \omega_i, \boldsymbol{\theta_i}) \right\} + \lambda \left(\sum_{i=1}^{c} P(\omega_i) - 1 \right) \tag{12}$$

We then calculate the partial derivative of equation (12) w.r.t. the generic mixing parameter $P(w_j)$ and set it equal to zero:

$$\frac{\partial L}{\partial P(\omega_j)} = \frac{\partial}{\partial P(\omega_j)} \sum_{k=1}^{N} \log \left\{ p(\boldsymbol{z^k} \mid \boldsymbol{\theta}) \right\} + \lambda \left(\sum_{i=1}^{c} P(\omega_i) - 1 \right)$$

$$= \sum_{k=1}^{N} \frac{1}{p(\boldsymbol{z^k} \mid \boldsymbol{\theta})} \frac{\partial}{\partial P(\omega_j)} \left\{ v^k \sum_{i=1}^{c} P(\omega_i) p(\boldsymbol{z^k} \mid \omega_i, \boldsymbol{\theta_i}) \right\} + \lambda$$

$$= \sum_{k=1}^{N} v^k \frac{p(\boldsymbol{z^k} \mid \omega_j, \boldsymbol{\theta_j})}{p(\boldsymbol{z^k} \mid \boldsymbol{\theta})} + \lambda$$

$$= \sum_{k=1}^{N} v^k \frac{P(\omega_j \mid \boldsymbol{z^k}, \boldsymbol{\theta})}{P(\omega_j)} + \lambda = 0 \tag{13}$$

where we used the equality

$$\frac{p(z^k \mid \omega_j, \boldsymbol{\theta}_j)}{p(z^k \mid \boldsymbol{\theta})} = \frac{P(\omega_j \mid z^k, \boldsymbol{\theta})}{P(\omega_j)} \tag{14}$$

given by Bayes theorem. Multiplying both sides of equation (13) by $P(w_j)$ and summing over j making use of the constraint $\sum_{i=1}^{c} P(w_i) = 1$, we obtain

$$
\begin{aligned}
\lambda &= -\sum_{j=1}^{c} P(\omega_j) \sum_{k=1}^{N} v^k \frac{P(\omega_j \mid z^k, \boldsymbol{\theta})}{P(\omega_j)} \\
&= -\sum_{k=1}^{N} v^k \sum_{j=1}^{c} \frac{P(\omega_j) P(\omega_j \mid z^k, \boldsymbol{\theta})}{P(\omega_j)} \\
&= -\sum_{k=1}^{N} v^k \sum_{j=1}^{c} P(\omega_j \mid z^k, \boldsymbol{\theta}) \\
&= -\sum_{k=1}^{N} v^k
\end{aligned}
\tag{15}
$$

Substitution of (15) into (13) gives

$$\frac{\sum_{k=1}^{N} v^k P(\omega_j \mid z^k, \boldsymbol{\theta})}{P(\omega_j)} = \sum_{k=1}^{N} v^k. \tag{16}$$

Finally, solving for $P(\omega_j)$:

$$P(\omega_j) = \frac{\sum_{k=1}^{N} v^k P(\omega_j \mid x^k, \boldsymbol{\theta})}{\sum_{k=1}^{N} v^k}. \tag{17}$$

Note that the derived maximum-likelihood estimation does not have a closed-form analytical solution. Then, following the classical EM approach [1] an iterative algorithm based on a gradient ascent procedure is exploited. Parameters $\boldsymbol{\theta}$ (i.e. $\boldsymbol{\mu}_j$, Σ_j and $P(\omega_j)$, for each Gaussian component j) are initialized arbitrarily (to this end, the *k-means* algorithm [1] is applied in this paper). Then, at each iteration, the E-step consists in computing $P(\omega_j \mid z^k, \boldsymbol{\theta})$ according to the current value of $\boldsymbol{\theta}$ and for each component j. In the M-step parameters are re-estimated using such values of $P(\omega_j \mid z^k, \boldsymbol{\theta})$ according to equations (10), (11) and (17) [1,3].

3   Extension to Multiple Hidden Layers

In this section we extend the training algorithm presented in the companion paper [2] to more-than-one hidden layer architectures. Algorithm 1 hands out the pseudo-code. The (outer) MLP is assumed to have L layers ($L - 1$ hidden layers with a layer-specific number of hidden units, and an output layer). The extension is rather straightforward. The basic idea requires an initialization via standard backpropagation (BP) as in [2]. The activation functions of the topmost hidden layer are basically trained as in the single-hidden-layer setup (called routine Train in the pseudo-code), via computation of the inverse of the output activation functions and backpropagation of the target outputs (referred to as routine BackpropagateTargets in Algorithm 1), as explained in [2]. Estimation of the corresponding GMM takes place according to the calculations given in Section 2. The weighted, joint pdf estimation of the GMM parameters is referred to as routine EstimateGMM in the pseudo-code. So far, the only novelty is that the input dataset (for estimation of the GMM and the training of the inner networks) is no longer obtained from the original input patters, but from the outputs yielded by the previous hidden layer (computed via routine FeedForward). Estimation of GMMs and training of inner MLPs within the lower hidden layers (down to the bottom-most) occur in an iterative fashion, following (i) a forward propagation of the original inputs up to the required layer, and (ii) a progressive backward propagation step of target outputs.

In Algorithm 1 actual inputs to l-th layer, for the k-th pattern are referred to as $x^k(l)$, while $\hat{x}^k(l)$ indicate the desired inputs (i.e. obtained through inversion of the activation functions for the L-th layer, and through MLP-inversion for the hidden layers, see below). The target outputs backpropagated at l-th layer is referred to as $o^k(l)$ [2]. $D_h(l) = \left\{ \left(x_h^k(l), o_h^k(l)\right), k = 1, \ldots, N \right\}$ denotes the training set for the h-th inner net in l-th layer, where $x_h^k(l)$ and $o_h^k(l)$ are the h-th entry of vector $x^k(l)$ and $o^k(l)$, respectively. Finally, $D_{GMM}(l) = \left\{ \left(x^k(l), o^k(l+1)\right), k = 1, \ldots, N \right\}$ is the dataset used for GMM estimation at l-th layer.

The only catch is the definition of suitable target outputs at a generic layer, starting from the outputs of the upper layer. This is accomplished by means of the so-called MLP *inversion* method [5], according to the calculations outlined in [4]. The method is conceptually simple: starting from a neural network which realizes a transformation $y = \phi(x, w)$ for a given (pre-trained) set of weights w, the MLP inversion principle prescribes the transformation of the input patterns x into new patters x' which better fit the target criterion function $C(\cdot)$. The update rule for creating x' follows the usual gradient descent approach, aimed at minimizing $C(\cdot)$ w.r.t. x (while the weights w are clamped to their original values). In summary, we let

$$x' = x - \eta \nabla_x C(\cdot) \tag{18}$$

whose explicit calculation is accomplished in a way similar to standard BP, $\eta \in \mathbb{R}^+$ being a learning rate. The routine realizing such an inversion scheme over a generic MLP is referred to as Invert() in the pseudo-code, and it has to be applied to all inner networks in the model.

Algorithm 1. Training multilayer networks with adaptive activation functions

Input: $D = \{(x^k, y^k), k = 1 \ldots N\}$
$L \leftarrow$ number of layers (except input layer)
for $l = 1$ **to** $L - 1$ **do**
    $m(l) \leftarrow$ number of units in l-th layer
end for
for $k = 1$ **to** N **do**
    $\widehat{x}^k(L) \leftarrow$ inverse of activation functions of L-th layer over y^k
    $o^k(L) = y^k$
end for
for $l = L - 1$ **down to** 1 **do**
    $D_{\text{GMM}}(l) \leftarrow \emptyset$
    for $h = 1$ **to** $m(l)$ **do**
        $D_h(l) \leftarrow \emptyset$
    end for
    for $k = 1$ **to** N **do**
        $x^k(l) \leftarrow$ FeedForward(x^k) up to l-th layer
        $D_{GMM}(l) \leftarrow D_{GMM}(l) \bigcup \{(x^k(l), o^k(l+1))\}$
    end for
    EstimateGMM over $D_{GMM}(l)$
    for $k = 1$ **to** N **do**
        $o^k(l) \leftarrow$ BackpropagateTargets$(\widehat{x}^k(l+1))$
        for $h = 1$ **to** $m(l)$ **do**
            $D_h(l) \leftarrow D_h(l) \bigcup \{(x_h^k(l), o_h^k(l))\}$
        end for
    end for
    for $h = 1$ **to** $m(l)$ **do**
        Train h-th inner net on $D_h(l)$
    end for
    for $k = 1$ **to** N **do**
        $\widehat{x}^k(l) \leftarrow$ Invert(inner networks)
    end for
end for

4   Demonstration

In this section we present a preliminary evaluation of the proposed model on a synthetic regression task. We generated piecewise functions defined over three intervals. In each interval the function is a mixture of basic functions, namely: a sinusoid multiplied by a quadratic function, a Gaussian mixture multiplied by a linear function and a cubic function. The order of the intervals is randomly generated for each piecewise function. Finally, random Gaussian noise was added to the function. The standard deviation of the noise is a random value varying between 0.01 and 0.05. The input and output range were normalized in $[-1, 1]$ and $[0, 1]$, respectively. The cardinality of the training, validation and test sets was 200, 100 and 200 patterns, respectively.

The model was evaluated making use of two common criteria, namely the mean squared error (MSE), that is $MSE = \frac{1}{N} \sum_{k=1}^{N} \left(y^k - \widetilde{y}^k\right)^2$, and the integrated squared error (ISE), defined as $ISE = \int_{\mathcal{I}} \left(f(x) - \widetilde{f}(x)\right)^2 dx$, where $\mathcal{I}$ is the interval where the x variable is defined. The integral was evaluated using Simpson method. To this end, the range of the input variable was divided into 1000 intervals. Figure 1a shows the original function (solid line) and the training set obtained by adding Gaussian noise (later used to train the outer network). We set $m = 2$ and then replaced each of the hidden units with the corresponding inner MLPs. The architecture of the latter ones was determined through a cross-validation procedure.

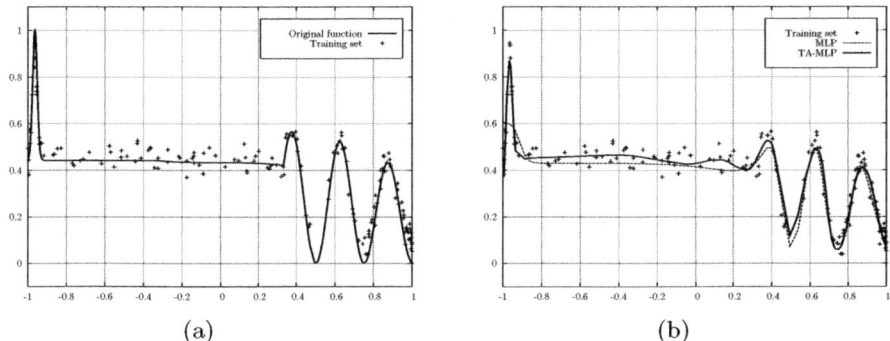

(a)   (b)

Fig. 1. (a) Synthetic function and training set - (b) Comparison between MLP and TA-MLP

Figures 2a and 2b show the training set, the probabilistic weights and the activation function learned by the two inner networks, respectively.

Table 1 shows the comparison between standard MLPs and TA-MLPs. For each network we indicate the total number of free parameters of the model, the number of hidden units (in case of TA-MLPs, the number of units of inner networks is also indicated in brackets), the MSE on both training and test sets, and the ISE. The best five results for both models are reported (in the order) in the table. The first row of the table shows that for a fixed number of free parameters (for both models) the TA-MLP achieves slightly better result than the MLP in terms of ISE and MSE, on both training and test sets. This confirms the algorithm is effective. Moreover, increased flexibility of the model over the training sample does not affect its generalization capabilities (i.e. proper, non-overfitting activation functions are actually learned). The subsequent rows show how the performance of TA-MLPs remains stable when the number of free parameters decreases. In traditional MLPs, on the other end, an increased number of free parameters does not entail a comparable improvement in terms of performance.

Figure 1b shows the approximations obtained with the best TA-MLP (solid line) and with the best standard MLP (dashed line), respectively, along with

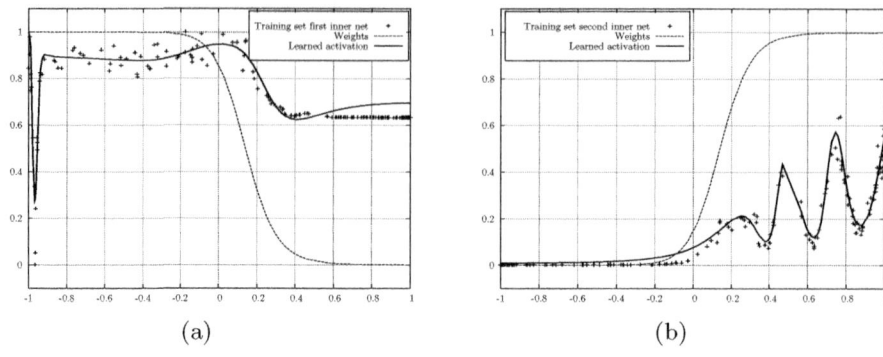

(a)   (b)

Fig. 2. Activation functions learned by (a) first inner network and (b) second inner network (solid lines), together with their training set (points) and their probabilistic weights (dashed lines)

Table 1. Comparison between MLPs and TA-MLPs

MLP					TA-MLP				
#Par	#Hid	MSEtrain	MSEtest	ISE	#Par	#Hid	MSEtrain	MSEtest	ISE
40	13	0.033	0.037	0.00137	40	2(5-6)	0.032	0.035	9.25e-4
49	16	0.046	0.052	0.00390	37	2(4-6)	0.033	0.035	9.42e-4
46	15	0.047	0.052	0.00393	34	2(4-5)	0.034	0.038	9.55e-4
34	11	0.047	0.052	0.00392	43	2(6-6)	0.033	0.035	9.67e-4
37	12	0.047	0.052	0.00399	37	2(5-5)	0.035	0.038	9.97e-4

the corresponding training sets. It is seen that modeling the first peak exhibited by the training data turned up infeasible via standard MLP, while using the TA-MLP the very peak turns out to be modeled suitably (via the inner network which focused on the corresponding, specific region). The activation functions learned by the two inner networks (problem-specific, and quite different from regular sigmoids) are shown in solid lines in figures 2a and 2b.

5   Conclusion

The paper developed an explicit, weighted maximum-likelihood solution to the problem of estimating the density functions (defined over the joint input/output space) associated wit the neurons of neural nets having adaptive activation functions. Combining the result with the generic training scheme introduced in the companion paper [2], a complete algorithm for this family of connectionist models emerges. The algorithm was extended to multi-layer architectures in a natural way. A preliminary experimental demonstration (over a synthetic regression task) proved the resulting approach being effective. It is seen that in the 1-hidden-layer scenario the overall machine can be described as a particular case of the traditional mixture of neural experts [6], having a novel training/gating policy. In the multi-layer setup this dual interpretation does not hold any longer, and we are faced with a novel, non-standard neural network (non fully-connected, and possibly having different depths along separate branches of its graphical structure), where the probabilistic measures associated with each adaptive neuron are defined over non-linearly transformed images of the original data. Efforts are currently focused toward (i) the definition of a robust, automatic technique for the selection of suitable, neuron-specific topologies for the inner MLPs (relying on the evaluation of a cross-validated log-likelihood criterion), as well as on (ii) a thorough experimental comparative analysis of the behavior of the proposed machine.

References

1. Duda, R.O., Hart, P.E., Stork, D.G.: Pattern Classification. Wiley (2001)
2. Castelli, I., Trentin, E.: Supervised and Unsupervised Co-Training of Adaptive Activation Functions in Neural Nets. In: Schwenker, F., Trentin, E. (eds.) PSL 2011. LNCS (LNAI), vol. 7081, pp. 52–61. Springer, Heidelberg (2012)
3. Bishop, C.M.: Pattern Recognition and Machine Learning. Information Science and Statistics. Springer, Heidelberg (2007)
4. Trentin, E., Gori, M.: Inversion-based nonlinear adaptation of noisy acoustic parameters for a neural/HMM speech recognizer. Neurocomputing 70(1-3), 398–408 (2006)
5. Linden, A., Kindermann, J.: Inversion of multilayer nets. In: Proc. of IJCNN 1989, Washington DC, pp. 425–430 (1989)
6. Hertz, J.A., Palmer, R.G., Krogh, A.: Introduction to the Theory of Neural Computation. Santa Fe Institute Studies in the Sciences of Complexity. Westview Press (1991)

Semi-Supervised Kernel Clustering with Sample-to-Cluster Weights

Stefan Faußer and Friedhelm Schwenker

Institute of Neural Information Processing, University of Ulm, 89069 Ulm, Germany
{stefan.fausser,friedhelm.Schwenker}@uni-ulm.de

Abstract. Collecting unlabelled data is often effortless while labelling them can be difficult. Either the amount of data is too large or samples cannot be assigned a specific class label with certainty. In semi-supervised clustering the aim is to set the cluster centres close to their label-matching samples and unlabelled samples. Kernel based clustering methods are known to improve the cluster results by clustering in feature space. In this paper we propose a semi-supervised kernel based clustering algorithm that minimizes convergently an error function with sample-to-cluster weights. These sample-to-cluster weights are set dependent on the class label, i.e. matching, not-matching or unlabelled. The algorithm is able to use many kernel based clustering methods although we suggest Kernel Fuzzy C-Means, Relational Neural Gas and Kernel K-Means. We evaluate empirically the performance of this algorithm on two real-life dataset, namely Steel Plates Faults and MiniBooNE.

1 Introduction

Given a mixture of labelled and unlabelled data the semi-supervised clustering methods aim to improve the cluster results with emphasize on the labelled data. In contrast the unsupervised clustering methods set the cluster centres independent on the data labels. Often only a small amount of labelled data is available as the task to label them is expensive. Most common the collected unlabelled data are too numerous to label it by hand or the human expert is unsure about the class labels of some of the data. Also problematic for fully supervised classification methods are ill-set class labels.

In the Learning-Vector Quantization (LVQ1) method (Kohonen 1997 [1]) the prototypes are attracted to the samples with the matching label and repulsed to the samples with the non-matching label by a modification of the learning rate. While this simple method delivers good results with data having many class labels it is known to not converge for samples with differing labels that overlap in the input space. Later on the Batch-LVQ method has been extended by fuzzy sample-to-cluster memberships (FSLVQ, Wu et al. 2003 [2]) but without considering the sample labels.

For a set of Must-Link (ML) and a set of Cannot-Link (CL) constraints on the data, the Hidden Markov Random Fields (HMRFs) algorithm for semi-supervised clustering (Basu et al. 2004 [3]) minimizes an error function with

F. Schwenker and E. Trentin (Eds.): PSL 2011, LNAI 7081, pp. 72–81, 2012.

penalty cost for violating the ML or CL constraints. While this method describes a sound way of semi-supervised clustering it needs at least one large matrix for the constraints and directly operates in the input space. In many cases samples are nonlinear distributed in input space and are easier to partition in feature space which makes kernel methods more useful. Two semi-supervised kernel based clustering algorithm one using seeded initialization sets of sample-to-cluster assignments (Seeded Kernel K-Means) and the other constraining samples from leaving their hard cluster assignment to their label-matching cluster (Constrained Kernel K-Means) have been shown to perform better than their non-kernel counterparts and HMRFs method (Yan et al. 2006 [5]). Later on it has been shown that the HMRFs method is similar to the Weighted Kernel K-Means algorithm for a linear kernel (Kulis et al. 2008 [4]). In the same article the authors introduced Semi-Supervised Kernel K-Means (SS-KERNEL-KMEANS) being basicly identical to Kernel K-Means but using a kernel matrix that embeds a constrained penalty matrix. In a very recent work Hu et al. (2010) ([6]) studied the semi-supervised kernel matrix learning (SS-KML) and tackled some of the performance problems by learning from a small kernel and propagating it into a larger-sized kernel. An interesting study that compares generally the semi-supervised classification of large-scale data can be found in Weston 2008 ([7]).

Our contribution in this paper is a novel semi-supervised kernel based clustering algorithm related to the Weighted Kernel K-Means algorithm with class labels. It is generalized to integrate multiple kernel based clustering methods. Specifically we integrate Kernel Fuzzy C-Means, Relational Neural Gas and Kernel K-Means (see section 2.1). This expands the algorithm by soft memberships. Further we broaden this algorithm to consider sample-to-cluster weights instead of only sample weights. We call this algorithm the Semi-Supervised Kernel Clustering with Sample-to-Cluster Weights (SKC). The objective function of this EM-style algorithm that it converges to is shown (see section 2). It is discussed how to set the sample-to-cluster weights based on the cluster and sample labels. In the end of section 3 it is discussed how to determine the cluster labels in a winner-takes-it all manner. Lastly in section 4 we evaluate empirically the performance of SKC on two real-life dataset, one being Steel Plates Faults with seven classes and the other being MiniBooNE with two classes.

2 Sample-to-Cluster Weighted Error Function

Assume that we want to partition N samples $x_1, x_2, \ldots, x_N$ with its assigned class labels $l_1, \ldots, l_N$ in K disjoint sets or cluster and each cluster has a representing prototype c_k and a class label l_k. The class labels for cluster and samples are both restricted to the same range of labels, i.e. $l_i, l_k \in \{0, 1, \ldots, M\}$ where $l_i = 0$ represents an unlabelled sample. We can then formulate the unnormalized sample-to-cluster weights $\beta_k(i)$:

$$\beta_k(i) = \begin{cases} 1 \text{ if } l_i = 0, \\ max(1, C\frac{|l_i=0|}{|l_i\neq0|}) \text{ if } l_i = l_k, \\ 0.01 \text{ if } l_i \neq l_k \end{cases} \tag{1}$$

For the case where all samples have a label $l_i \neq 0$ the sample-to-cluster weights $\beta_k(i)$ rewards samples whose label matches to the cluster label by 1 and penalizes the samples with not-matching labels by 0.01. If there are unlabelled samples then the summed weights of the labelled samples differ from the summed weights of the unlabelled samples by the factor $C \geq 1$. Further the normalized sample-to-cluster weights $\alpha_k(i)$ are:

$$\alpha_k(i) = \frac{\beta_k(i)}{\sum_{j=1}^{N} \beta_k(j)} \tag{2}$$

Compared to the unnormalized sample-to-cluster-weights, the normalized sample-to-cluster weights $\alpha_k(i)$ all sum to unity for each cluster k, i.e. $\alpha_k(i) > 0, \forall i$ and $\sum_{i=1}^{N} \alpha_k(i) = 1$. Therefore the sample labels only effect the weights within the same cluster. Having defined the normalized sample-to-cluster weights we can now introduce the sample-to-cluster weighted quantization error as follows:

$$E(c_k) = \sum_{k=1}^{K} \sum_{i=1}^{N} f_k(i)\alpha_k(i)d(c_k, x_i) \tag{3}$$

where $f_k(i)$ is a hard or bounded soft assignment of sample i to cluster k and $d(c_k, x_i)$ is the distance between sample x_i and cluster prototype c_k. If $d(c_k, x_i)$ is measured by the euclidean distance, $\alpha_k(i) = 1/N, \forall i, \forall k$ and $f_k(i)$ is a hard assignment then $E(c_k)$ is the exact quantization error that K-Means minimizes. The steps of the K-Means algorithm are 1. updating $f_k(i)$: hard assign samples to clusters based on their distance to their nearest prototypes and 2. updating c_k: move prototypes to their cluster centres. Both steps are repeated until the prototypes converges locally. If we consider the euclidean distance then we can derive the positions of the prototypes c_k by gradient descent, i.e. calculating $\frac{\partial E(c_k)}{\partial c_k} = 0$. This results in the following general function to calculate the current prototype positions given the sample-to-cluster assignments $f_k(i)$ and the sample-to-cluster weights $\alpha_k(i)$:

$$c_k = \frac{\sum_{i=1}^{N} \alpha_k(i)f_k(i)x_i}{\sum_{i=1}^{N} \alpha_k(i)f_k(i)}$$

Analyzing this formula a cluster centre is more attracted to samples with matching class labels, less attracted to unlabelled samples and almost unaffected to samples with opposing class labels. Now suppose that we would transform all samples $x_i \in X$ and prototypes $c_k \in X$ to a feature space using the mapping function $\phi : X \rightarrow \mathbb{F}$ that maps X from input space to a possible high-dimensional feature space $\mathbb{F}$. Note that X can be an arbitrary set, e.g. $\mathbb{R}^n$. This would allow us to calculate the prototypes in feature space. The benefit is that it can be easier to partition the samples in the feature space than in the origin input space. Unfortunately such mapping functions are costly and often unknown. Still we can calculate the distance between such (theoretically) transformed samples using a positive-definite and symmetric kernel $\kappa(x_i, x_j)$ and applying the kernel trick, i.e. define the prototypes as linear combinations of existing transformed samples. Considering the prior defined sample-to-cluster weights we can then set up the new sample-to-cluster-weighted distance function $d_{weighted}(\phi(c_k), \phi(x_i))$ in feature space. The distance function $d_{weighted}(\phi(c_k), \phi(x_i))$ can be written as:

$$d_{weighted}(\phi(c_k), \phi(x_i)) = ||\phi(x_i) - \frac{\sum_{j=1}^{N} f_k^\phi(j) \alpha_k(j) \phi(x_j)}{\sum_{j=1}^{N} f_k^\phi(j) \alpha_k(j)}||^2 \tag{4}$$

$$= \langle \phi(x_i), \phi(x_i) \rangle - 2 \left\langle \phi(x_i), \frac{\sum_{j=1}^{N} f_k^\phi(j) \alpha_k(j) \phi(x_j)}{\sum_{j=1}^{N} f_k^\phi(j) \alpha_k(j)} \right\rangle +$$

$$\left\langle \frac{\sum_{j=1}^{N} f_k^\phi(j) \alpha_k(j) \phi(x_j)}{\sum_{j=1}^{N} f_k^\phi(j) \alpha_k(j)}, \frac{\sum_{j=1}^{N} f_k^\phi(j) \alpha_k(j) \phi(x_j)}{\sum_{j=1}^{N} f_k^\phi(j) \alpha_k(j)} \right\rangle$$

$$= \kappa(x_i, x_i) - \frac{2 \sum_{j=1}^{N} f_k^\phi(j) \alpha_k(j) \kappa(x_i, x_j)}{\sum_{j=1}^{N} f_k^\phi(j) \alpha_k(j)} +$$

$$\frac{\sum_{j=1}^{N} \sum_{l=1}^{N} f_k^\phi(j) f_k^\phi(l) \alpha_k(j) \alpha_k(l) \kappa(x_j, x_l)}{[\sum_{j=1}^{N} f_k^\phi(j) \alpha_k(j)]^2}$$

While minimizing the quantization error in input space (e.g. standard K-Means) can be achieved by repeatedly moving the cluster prototypes this is not directly possible in feature space. Instead the aim is to set the cluster assignments to such values that the quantization error (see equation (3)) will be minimized. This can be done by iteratively updating the assignments by calculating and comparing the distances. Note that an update of the cluster assignments f also implicitly changes the positions of the prototypes in feature space.

2.1   Weighted Kernel Based Methods for Clustering

Having defined the sample-to-cluster weighted distance function (see equation (4)) we can now use this distance measure in feature space and define the sample-to-cluster assignment functions for the most common divisive kernel clustering methods. Selecting one of the clustering methods is as simple as selecting one of the following assignment functions f. This is possible as these kernel clustering methods only differ by their sample-to-cluster assignment function f. As for Sample-to-Cluster Weighted Kernel K-Means the assignment update step is:

$$f_k^\phi(i) = \begin{cases} 1 \text{ if} & d_{weighted}(\phi(c_k), \phi(x_i)) < d_{weighted}(\phi(c_m), \phi(x_i)), \\ & m = 1, \dots, K, m \neq k \\ 0 \text{ else} \end{cases} \tag{5}$$

For Sample-to-Cluster Weighted Relational Neural Gas (for the basic method see Hammer et al. 2007 [9]) the assignment update step is:

$$f_k^\phi(i) = exp\left(\frac{-rank(\phi(c_k), \phi(x_i))}{\lambda}\right) \tag{6}$$

where $rank(\phi(c_k), \phi(x_i)) = |\{\phi(c_m) \mid d_{weighted}(\phi(c_m), \phi(x_i)) < d_{weighted}(\phi(c_k), \phi(x_i)), m = 1, \dots, K, m \neq k\}| \in \{0, \dots, K-1\}$. Lastly the assignment update steps for Sample-to-Cluster Weighted Kernel Fuzzy C-Means (for the basic method see Zhang et al. 2002 [8]) gets:

$$f_k^\phi(i) = \frac{1}{\sum_{n=1}^{K}\left[\frac{d_{weighted}(\phi(c_k), \phi(x_i))}{d_{weighted}(\phi(c_n), \phi(x_i))}^{\frac{2}{m-1}}\right]} \tag{7}$$

Comparing the three algorithms, Kernel Fuzzy C-Means and Relational Neural Gas are more insensitive to initializations as both algorithms update their indirectly defined prototypes not only by their greedy winner samples but also by other samples determined through neighborhood size λ (Relational Neural Gas) or fuzzifier m (Kernel Fuzzy C-Means). For Relational Neural Gas this neighborhood size reduces during the increasing iteration steps t, e.g. by $N/(2 \cdot t)$. The Kernel Fuzzy C-Means algorithm however uses and returns soft assignments, i.e. gives possibly more information on the data. On the other hand if fuzzifier $m \rightarrow 1$ then this algorithms behaves exactly like Kernel K-Means.

3   Semi-Supervised Kernel Clustering with Sample-to-Cluster Weights

Given the samples x_i with sample labels l_i, a kernel function κ, the number of cluster K, their assigned cluster labels l_k and the kernel clustering method, i.e. the method to calculate the sample-to-cluster assigments $f_k(i)$ (see section 2.1) we can formulate the Semi-Supervised Kernel Clustering with Sample-to-Cluster Weights (SKC) algorithm in algorithm 1. The algorithm has converged to a local minima if the cluster assignments and therefore also the cluster prototypes do not change anymore or the sum of the differences between the old and new cluster assignments gets below a certain threshold. The output of this algorithm are then the assignments f that determines the hard (Relational Neural Gas, Kernel K-Means) or soft (Kernel Fuzzy C-Means) assignments of the given samples to K cluster. Further the distances of samples that have not been deployed in the above algorithm can be calculated to the resulting prototypes using equation (4).

Algorithm 1. Semi-Supervised Kernel Clustering with Sample-to-Cluster Weights (SKC) algorithm

Input: samples $x_i \in X^N$ with labels $l_i \in \{0, \ldots, M\}$, cluster labels l_k, kernel function κ, number of cluster K and kernel clustering method (Kernel K-Means, Kernel Fuzzy C-Means, Relational Neural Gas)

Output: final cluster assignments f

    Arbitrary set sample-to-cluster-assignments f

    Calculate sample-to-cluster weights using equation (2)

    repeat

        Calculate $d_{weighted}(\phi(c_k), \phi(x_i))$ of the samples x_i to the cluster centres c_k in feature space using equation (4)

        Calculate f given the kernel clustering method

    until convergence

This is possible as the resulting prototypes are defined in feature space as a linear combination of the used samples, the final cluster assignments f and the calculated sample-to-cluster weights α. Therefore this algorithm can be used to 1. semi-supervised cluster data with all known labels or 2. semi-supervised cluster data with some known labels and 3. use the resulting prototypes to detect the labels of new samples. The given cluster labels can be estimated based on the amounts of known sample labels, i.e. $\sim K \cdot \frac{|l_i|l_i \neq 0|}{\sum_j |l_j|l_j \neq 0|}$ cluster with the label $l_k = l_i$.

Although it is possible in the SKC algorithm to only know a small amount of sample labels, it might impose some problems. Suppose to know only a small amount of sample labels, i.e. most sample labels are $l_i = 0$. Now estimating the real amount of sample labels and therefore distributing the class labels to the cluster is no easy task. On the other hand we can roughly guess the cluster label l_k by the current given hard assigned samples:

$$l_k = \operatorname*{argmax}_{j=1,\ldots,M}(|l_i = j, x_i \in c_k|) \tag{8}$$

In this winner-takes-it-all method the cluster label $l_k \in \{1, \ldots, M\}$ gets the label that currently the most hard assigned samples have. Also this equation directly maximizes the Purity (see section 4). Integrated in the prior defined algorithm SKC the class-label winner-detection is performed after calculating the sample-to-cluster assignments f (after step 2). Further the sample-to-cluster weights (equation (2)) have to be calculated each time the cluster label changes. The cluster-label need not be initialized but can be determined by initial f using this method.

4   Experiments and Results

To evaluate empirically the performance of our SKC method we have tested it on two real-life dataset (Steel Plates Faults and MiniBooNE). We compared the performance to the de-facto standard Constrained Kernel K-Means with Seeded Sets (CKKMEANS, Yan et al. 2006 [5]) and the basic kernel clustering methods without labels. For the smaller dataset (Steel Plates Faults) we either randomly removed some of the sample labels (70 percent) or used all sample labels for the semi-supervised clustering and then calculated the summed Total-Within Cluster Variation with the resulting cluster centres (TWCV, see equation (3) but with hard sample-to-cluster assignments $f_k(i)$ and $\alpha_k(i) = 1/N, \forall i, \forall k$). Further we show the external cluster validation measures Purity and Normalized Mutual Information (NMI) with all known sample labels. The Purity is the summed number of highest sample labels in each cluster divided by the number of samples N, i.e. $Purity = \frac{1}{N} \sum_{k=1}^{K} max_{l_i, x_i \in c_k} |l_i|$. If the number of clusters are higher than the number of class labels then a better criteria can be the NMI: $nMI(C, K) = \frac{I(C,K)}{\sqrt{H(C) \cdot H(K)}}$ where $I(C, K)$ measures the mutual information between the classes C and the cluster K normalized by the entropy of classes C ($H(C)$) and the entropy of cluster K ($H(K)$). Both Purity and NMI range between 0 and 1, higher values are better. For the large-scale dataset MiniBooNE we chose the first 1000 samples for each class and calculated additionally the TWCV, Purity and NMI with all $\sim 130,000$ samples. As for the parameter we set λ for RNG to linear decrease from $N/2$ to 0.01, fuzzifier m for KFCMEANS to 1.25 and factor $C = 2$ (see equation (1)) for all experiments.

4.1   Steel Plates Faults Dataset

The Steel Plates Faults Dataset (UCI repository [10]) consists of $1,941$ samples $\in \mathbb{R}^{27}$ with seven class labels. It describes the visual image of a steel plate by luminosity, orientation, edges etc. where each sample has one of seven faults, i.e. pastry, Z-scratch, K-scratch, stains, dirtiness, bumps and other. We normalized the samples to zero mean and unit variance and used the RBF-Kernel to calculate the similarities between the samples:

$$\kappa(x_i, x_j) = exp(-\frac{||x_i - x_j||^2}{2\sigma^2}) \tag{9}$$

Several values for parameter σ were tested with basic Kernel K-Means for $K = 14$ cluster to maximize Purity and the NMI value and to minimize the TWCV. Therefore any improvements to the KKMEANS results are significant. Parameter $\sigma = 4$. The results for SKC with given cluster labels as well as cluster-label winner detection can be seen in table 1. Our SKC algorithm combined with KKMEANS, KFCMEANS or RNG are best both in terms of external cluster validation (Purity and NMI) as well as the quantization error (TWCV) compared to the standard kernel methods (KKMEANS, KFCMEANS and RNG) as well as the Constrained Kernel K-Means (CKKMEANS) method. The Purity could be increased from 58.8 to 64.5 percent (SKC-RNG). Most interesting the CKKMEANS algorithm that we compare to has a good Purity and NMI value but too high TWCV for all label samples (label amount 1.0). For this case the algorithm is so constrained that it simply cannot cluster but only return the given seed set. However our SKC algorithm has only slightly higher quantization errors with all sample labels than with some or no sample labels, i.e. still produces good clusterings in terms of the TWCV.

Table 1. Cluster validations on Steel Plates Faults dataset. The values are averaged over 30 testruns (with small variances, omitted). All methods had to partition the samples in $K = 14$ cluster. Second column (label amount) describes the factor of known sample labels. The third column describes if cluster-labels are predefined (n) or detected during the algorithm (y). External cluster validation measures are given in the fourth and fifth column. The internal cluster validation measure (TWCV) is given in the last column.

	fac.	det.	Purity	NMI	TWCV
KKMEANS	-	-	0.593	0.3357	761
KFCMEANS	-	-	0.602	0.3412	762
RNG	-	-	0.588	0.3321	761
CKKMEANS	0.3	n	0.618	0.3534	798
CKKMEANS	1.0	n	0.647	0.37	1001
SKC-KKMEANS	0.3	y	0.615	0.3501	796
SKC-KFCMEANS	0.3	y	0.619	0.3526	789
SKC-RNG	0.3	y	0.617	0.3507	777
SKC-KKMEANS	0.3	n	0.617	0.3549	794
SKC-KFCMEANS	0.3	n	0.623	0.3528	778
SKC-RNG	0.3	n	0.627	0.3538	777
SKC-KKMEANS	1.0	n	0.646	0.37	808
SKC-KFCMEANS	1.0	n	0.645	0.3688	809
SKC-RNG	1.0	n	0.645	0.3672	802

4.2 MiniBooNE Dataset

In the MiniBooNE Dataset from the UCI repository (see [10]), electron neutrinos (signal) have to be divided from muon neutrinos (background). It is a large-scale dataset with $130,065$ samples $\in \mathbb{R}^{50}$ and two class labels. For the (semi-supervised) clustering we chose the first 1000 samples of each class, i.e. clustered 2000 of the samples. We normalized the 50 particle id attributes to zero mean and unit variance and used the RBF-Kernel (see equation (9)) with parameter $\sigma = 3$ for the similarity measure. The result of the clustering algorithm can be seen in table 2. This time we compared additionally the Purity, NMI values and TWCV of all samples to the resulting prototypes. If 30 percent of the sample labels are known then our SKC algorithm performs best versus all other methods (KKMEANS, KFCMEANS, RNG and CKKMEANS) in terms of lower TWCV and higher NMI and Purity values. However for all known sample labels the CKKMEANS algorithm performs slightly better than SKC-KKMEANS for Purity and NMI but has much worse TWCV. Similar results have been observed for the Steel Plates Faults dataset (see section 4.1) where we have concluded that the CKKMEANS algorithm is too constrained for all known samples and unable to cluster. The SKC-RNG algorithm however has about the same NMI and Purity values and much better TWCV than CKKMEANS. The NMI for SKC-RNG could be increased from 14.02 to 17.58 percent (0.3 labels) respectively to 18.82 percent (1.0 labels).

Table 2. Cluster validations on MiniBooNE dataset. The values are averaged over 30 testruns (with small variances, omitted). All methods had to partition the samples in $K = 6$ cluster. Second column (label amount) describes the factor of known sample labels. The third column describes if cluster-labels are predefined (n) or detected during the algorithm (y). The external cluster validation measures are given for the clustered samples (fourth and fifth column) and for all samples (sixth and seventh column). The internal cluster validation measure (TWCV) is given in the last column for all samples.

	fac.	det.	Purity	NMI	Purity all	NMI all	TWCV all
KKMEANS	-	-	0.761	0.1893	0.78	0.1626	24815
KFCMEANS	-	-	0.759	0.1854	0.793	0.1591	24754
RNG	-	-	0.737	0.1626	0.783	0.1402	24740
CKKMEANS	0.3	n	0.768	0.1914	0.786	0.164	25830
CKKMEANS	1.0	n	0.796	0.2069	0.838	0.1856	37034
SKC-KKMEANS	0.3	y	0.779	0.1978	0.793	0.1728	25013
SKC-KFCMEANS	0.3	y	0.777	0.1971	0.8	0.171	24892
SKC-RNG	0.3	y	0.789	0.2057	0.796	0.1758	24831
SKC-KKMEANS	0.3	n	0.771	0.191	0.794	0.1653	24894
SKC-KFCMEANS	0.3	n	0.755	0.1786	0.801	0.1552	24789
SKC-RNG	0.3	n	0.774	0.1936	0.801	0.1648	24629
SKC-KKMEANS	1.0	y	0.788	0.2051	0.81	0.177	24964
SKC-KFCMEANS	1.0	y	0.781	0.1977	0.81	0.1746	25071
SKC-RNG	1.0	y	0.8	0.2185	0.8	0.1882	25074

5 Conclusion

We have defined an objective function based on the quantization error and with sample-to-cluster weights. Using this error function we derived the semi-supervised kernel based clustering algorithm that integrates multiple kernel methods specifically Kernel K-Means, Kernel Fuzzy C-Means and Relational Neural Gas. We have shown how to set these sample-to-cluster weights and how these weights influence the positions of the cluster centres. Further we extended the algorithm to detect the cluster-labels in a winner-takes-it-all manner. We have performed experiments on two real-life dataset. For the large-scale dataset (MiniBooNE) we clustered with 2000 samples and have shown the TWCV, Purity and NMI values with the resulting cluster centres and all ($\sim 130,000$) samples. We have shown empirically for both dataset that our algorithm (SKC) has better TWCV, Purity and NMI values than the basic kernel clustering methods (KKMEANS, KFCMEANS and RNG) and the Constrained Kernel K-Means (CKKMEANS) method. What remains of interest is how the SKC algorithm performs with kernels that include sample label informations or sample-to-sample constraints.

References

1. Kohonen, T.: Self-organizing maps. Springer, Heidelberg (1997)
2. Wu, K.-L., Yang, M.-S.: A fuzzy-soft learning vector quantization. Neurocomputing 55(3-4), 681–697 (2003)
3. Basu, S., Bilenko, M., Mooney, R.J.: A Probabilistic Framework for Semi-Supervised Clustering. In: Proceedings of the 10th ACM SIGKDD International Conference on Knowledge Discovery and Data-Mining, pp. 59–68 (2004)
4. Kulis, B., Basu, S., Dhillon, I., Mooney, R.: Semi-supervised Graph Clustering: A Kernel Approach. In: Proceedings of the 25th International Conference on Machine Learning, vol. 74(1), pp. 1–22 (2008)
5. Yan, B., Domeniconi, C.: Exploration of Different Constraints and Query Methods with Kernel-based Semi-Supervised Clustering. In: IEEE International Conference on Systems, Man and Cybernetics, SMC 2006, pp. 829–834 (2006)
6. Hu, E., Chen, S., Zhang, D., Yin, X.: Semisupervised Kernel Matrix Learning by Kernel Propagation. IEEE Transactions on Neural Networks 21(11) (2010)
7. Weston, J.: Large-Scale Semi-Supervised Learning. In: Proceedings of NATO Advanced Study Institute on Mining Massive Data Sets for Security, vol. 19, pp. 62–75 (2008)
8. Zhang, D.Q., Chen, S.C.: Fuzzy clustering using kernel methods. In: International Conference of Control and Automatation, ICCA 2002, pp. 123–128 (2002)
9. Hammer, B., Hasenfuss, A.: Relational Neural Gas. In: Hertzberg, J., Beetz, M., Englert, R. (eds.) KI 2007. LNCS (LNAI), vol. 4667, pp. 190–204. Springer, Heidelberg (2007)
10. Frank, A., Asuncion, A.: UCI Machine Learning Repository, University of California, School of Information and Computer Sciences, Irvine (2010), http://archive.ics.uci.edu/ml

Homeokinetic Reinforcement Learning

Simón C. Smith and J. Michael Herrmann

Institute of Perception, Action and Behaviour, School of Informatics,
The University of Edinburgh, 10 Crichton St, Edinburgh, EH8 9AB, U.K.
{artificialsimon,michael.herrmann}@ed.ac.uk

Abstract. In order to find a control policy for an autonomous robot by reinforcement learning, the utility of a behaviour can be revealed locally through a modulation of the motor command by probing actions. For robots with many degrees of freedom, this type of exploration becomes inefficient such that it is an interesting option to use an auxiliary controller for the selection of promising probing actions. We suggest here to optimise the exploratory modulation by a self-organising controller. The approach is illustrated by two control tasks, namely swing-up of a pendulum and walking in a simulated hexapod. The results imply that the homeokinetic approach is beneficial for high complexity problems.

1 Introduction

Reinforcement Learning, discrete [1,12,13] as well as continuous [5], aims at solving dynamical optimisation problems. For this purpose a utility function and/or a control policy is constructed. Optimal performance can be reached asymptotically under certain conditions. However, because often Markovian state transitions and slow decay of the learning rate cannot be asserted in practical problems, only suboptimal solutions are found.

Additionally, in high dimensions, the exploration of the state space is time consuming. Gradient-based reinforcement learning can speed-up the optimisation process, but is prone to local optima, and if the gradient is not known then probing actions must be used in order to obtain gradient information. High-frequency probing [14] tests two alternative actions virtually at the same time which seems appropriate for an autonomous agent which may not be able to apply different actions in the same state. In addition, the set-up of the probing actions requires some domain knowledge and becomes cumbersome in high dimensions. Furthermore, what priorities should be used when sequentially probing the manifold of behaviours in robots with many degrees of freedom? A robot with a reinforcement learning (RL) controller is biased to keep trying the path that is expected to give him the best reward in the future, thus seemingly non rewarding nearby states are less likely to be explored.

We propose to use an auxiliary algorithm that learns to probe the system. For this purpose we will not follow the gradient of the utility function, but will aim at maximising the learning success achieved by the probing actions. This will help to obtain a more reliable representation of the utility function in shorter time,

F. Schwenker and E. Trentin (Eds.): PSL 2011, LNAI 7081, pp. 82–91, 2012.

while the reinforcement learning component will be responsible for the actual increase of the expected reward.

The probing algorithm relies on a self-organising (SO) control paradigm described in Ref. [10]. The SO controller generates motors signals based on estimated next sensor values. This signal will be used by the reinforcement learning controller as exploration mode and to update the parameters values in an actor-critic configuration. A representation of the world and the controller are modelled and updated based on the time loop error, see [3,4]. In Fig. 1 a scheme of the architecture can be seen, the reinforcement learning controller generates a motor signal u_t given the actual states x_t and the exploratory signal n_t provided by the SO controller. Given the actual motor signal and the actual state the model predicts the next sensor input which is used to calculate the time loop error and to update the SO controller. It is essential to the approach followed here that the full loop through the environment is monitored by the robot. This loop can be represented by a map of previous to new sensor values, but as well also as a map from previous to new motor commands. The latter case is actually more convenient if as often the dimensionality of the motor space is lower than that of the sensor space.

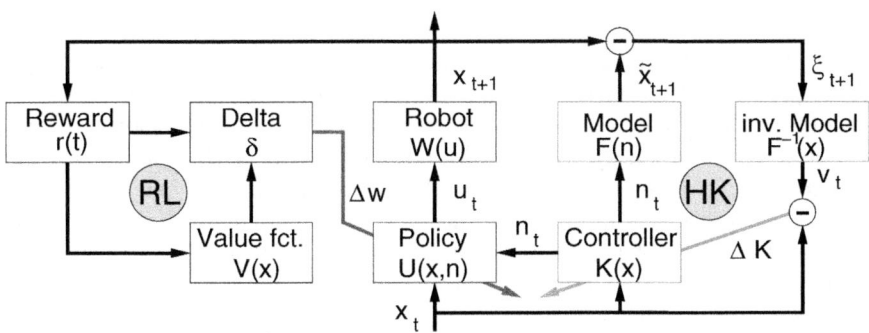

Fig. 1. Architecture of the sensorimotor loop (RL: Reinforcement learning controller, HK: Homeokinetic controller; for other symbols see Section 2)

We present a comparison of our approach with a standard version of continuous reinforcement learning [5] in the low dimensional task of swinging up a pendulum with limited torque and in an hexapod robot with twelve degrees of freedom where walking speed is to be optimised.

2   Reinforcement Learning in Continuous Space and Time

Following [5], the control command is given by

$$u_t = U_t \left(x_t \right) = s \left(A \left(x_t; w^A \right) + \sigma \mathbf{n}_t \right),$$ (1)

where s is the output function, $\mathbf{n}$ is a probing input of strength σ and

$$A\left(x_t; w^A\right) = N\left(x\right) \sum_i w_i^A \exp\left(-\frac{\|x_t - \mu_i\|^2}{2\rho_i^2}\right) \tag{2}$$

is a policy function that depends on parameters w^A (ρ_i and μ_i are assumed to be fixed). The factor $N\left(x\right) = \left(\sum_i \exp\left(-\frac{\|x_t - \mu_i\|^2}{2\rho_i^2}\right)\right)^{-1}$ normalises the actor output. The parameters w^A are updated according to

$$\Delta w_i^A = \eta^A \delta_t \mathbf{n}_t \frac{\partial A\left(x_t; w^A\right)}{\partial w_i^A}, \tag{3}$$

where η^A is a learning rate.

While the last term in (3) is easily obtained from (2), the essential part of this learning rule includes the correlation of the probing input $\mathbf{n}$ and the delta error

$$\delta_t = r_t - \frac{1}{\tau} V_t + \dot{V}_t \tag{4}$$

The utility function V is represented by another parametrised function which is simultaneously updated.

There are various ways of choosing the probing excitation of the robot control in Eq. 1. Gullapalli [7] suggested to use noise while others [14,2] have proposed high-frequency oscillatory modulations of the motor command. Our experiments confirm that the type of the probe does not matter in low-dimensional problems. For robots with many degrees of freedom, the dynamics of the correlation among the degrees of freedom of the controlled system becomes crucial such that the choice of the probing stimulus becomes non-trivial. In high-dimensional problems it is obviously not possible to test all actions in all states infinitely often as it would be required in discrete reinforcement learning algorithms. Also for continuous algorithms orienting the exploration to promising directions is essential. We propose to use an approach in the present context that we have previously developed in a different setting [8].

3   Learning in Motor Space

Instead of using noisy probing, we propose to modulate the motor command (1)

$$u_t = s\left(A\left(x_t; w^A\right) + \sigma K\left(x_t\right)\right) \tag{5}$$

by an exploratory controller

$$K\left(x_t\right) = g\left(Cx_t + C_0\right). \tag{6}$$

This controller receives the current sensory input vector x_t and determines the direction of exploration in dependence on the multidimensional parameters $C \in \mathbb{R}^{m \times n}$ and $C_0 \in \mathbb{R}^m$ and a further nonlinear function g. In order to adapt the parameters C and C_0, the new sensory inputs are compared with a prediction $\hat{x}$ by a world model M based on previous inputs or outputs. For simplicity we use a linear predictor that uses only the motor commands (5) and receives thus information about previous inputs only indirectly.

$$\hat{x}_{t+1} = M\left(u_t\right) = Du_t + D_0 \tag{7}$$

The comparison of the corresponding sensory input x_{t+1} and its estimate by the internal model $\hat{x}_{t+1}$ results in the prediction error $\xi_{t+1} = \hat{x}_{t+1} - x_{t+1}$ which is a vector in the perceptual space.

In order to formulate a learning rule for the exploratory controller (7) we will follow the procedure in Ref. [8] and express the error in the motor space which can be achieved by defining a transformed error η_t via

$$M\left(u_t\right) + \xi_{t+1} = M\left(u_t + \eta_t\right). \tag{8}$$

Because $M\left(u_t\right) + \xi_{t+1} = x_{t+1}$, the motor error η can be interpreted as the control correction required to compensate the inaccuracy of the model M. η is a retrospective error that can be determined only after the event of receiving the new stimulus x_{t+1}. Nevertheless, minimisation of η is a relevant goal for the adaptation of the system. The definition (8) is implicit and may be empty which calls for the use of a regularised inverse of M to explicitly obtain an approximation of η. Practically, Eq. 8 is transformed into a motor level error exploiting the assumed linearity of the model (7),

$$\eta_t = M'^{+}\xi_{t+1}, \tag{9}$$

where M'^{+} is the pseudo-inverse of the derivative of the model (7), i.e. the pseudoinverse of D in Eq. 7. In analogy to Ref. [3] this defines a homeokinetic error function in the motor space

$$E_t = \eta_t^{\top}\left(J_t J_t^{\top}\right)^{-1}\eta_t \tag{10}$$

where J is the Jacobian of the sensorimotor loop, see below. We are going to perform a gradient descent with respect to this error function in order to adapt the parameters of the controller (6).

To calculate the Jacobian, we use the derivatives $M_u' = D$ and $U_x' = s' \circ \left(\frac{\partial A}{\partial x} + \sigma g' \circ C\right)$ such that we find from $J_t = \phi_u' = U_x'\left(x_t\right)M_u'\left(x_{t-1}; u_{t-1}\right)$[1]

$$J_t = \left(\frac{\partial A}{\partial x} + \sigma g_t' \circ C_t\right)D_t.$$

This gives rise to the following formulation of the shift ν, i.e. the change in motor command that would have been required to correctly predict the following motor command, namely

$$\nu_{t-1} = J_t^{-1}\eta_t$$

---

[1] The dependence $\frac{dx}{du}$ of the new sensory input on the motor command is approximated here based on the assumption of a correct model (certainty equivalence). The symbol $\circ$ denotes component-wise multiplication.

While the above interpretation (9) of η as retrospective error connects sensor and motor space, we have here a connection between the two points in time within the motor space that reflects the dynamical properties of the full sensorimotor loop. The error function (10) becomes thus simply

$$E_t = \nu_{t-1}^\top \nu_{t-1}$$

which lead to a convenient update rule of the controller matrix C. Omitting the time indices we find

$$\frac{1}{\varepsilon_C} \Delta C = -\frac{\partial E}{\partial C} = -2\nu^\top \frac{\partial \nu}{\partial C} = 2\nu^\top J^{-1} \frac{\partial J}{\partial C} J^{-1}\eta - 2\nu^\top J^{-1} \frac{\partial \eta}{\partial C}$$

using the rule $\frac{\partial Y^{-1}}{\partial X} = -Y^{-1}\frac{\partial Y}{\partial X}Y^{-1}$. The derivative $\frac{\partial \eta}{\partial C}$ cannot be determined, because we have no information of the dependence of the prediction error on the controller parameters, therefore we set $\frac{\partial \eta}{\partial C} = 0$ and are left with

$$\frac{1}{\varepsilon_C} \Delta C = 2\nu^\top J^{-1} \frac{\partial J}{\partial C} J^{-1}\eta = 2\nu^\top J^{-1} \frac{\partial J}{\partial C}\nu$$

where

$$\frac{\partial J_t}{\partial C} = \frac{\partial}{\partial C}\left(\frac{\partial A}{\partial x} + \sigma g_t' \circ C_t\right) D_t.$$

We may ignore the effect of the controller on the sensitivity of the actor in the reinforcement learning component, i.e. set $\frac{\partial}{\partial C}\frac{\partial A}{\partial x} = 0$. We may also assume that the details of the actor are not specified by the reward but will follow essentially the homeokinetic control. In this case the term $\frac{\partial}{\partial C}\frac{\partial A}{\partial x}$ is parallel to the remainder and the resulting numerical factor can be absorbed into the learning rate. We have thus arrived at essentially the same learning rule as in Ref. [8],

$$\frac{1}{\varepsilon_C} \Delta C = \chi \left(D\nu\right)^\top - \chi^\top \frac{\partial g'^{-1} \circ \eta}{\partial C},$$

which, however, is to be evaluated at the controller with the reinforcement learning component.

Inserting the correct time indexes we obtain

$$\frac{1}{\varepsilon_C} \Delta C = \chi_{t-1} \left(D_t\nu_{t-2}\right)^\top - 2\left(\chi_{t-1} \circ g_{t-2} \circ \left(g_{t-2}'\right)^{-1} \circ \eta_{t-1}\right) x_{t-2}^\top$$

with $\chi_{t-1} = \left(R_t^\top\right)^{-1}\nu_{t-2}$. The update rule for C_0 can be found analogously,

$$\frac{1}{\varepsilon_C} \Delta C_{0,t} = -2\left(\chi_{t-1} \circ g_{t-2} \circ \left(g_{t-2}'\right)^{-1} \circ \eta_{t-1}\right)$$

4 Homeokinetic Reinforcement Learning: Experiments

In order to test our approach two nonlinear control task were implemented. The first one is the pendulum swing-up task (number of sensors $n = 2$, number of motors $m = 1$) where a pendulum has to be brought to the upright position to obtain the maximum reward. The second task consist in teaching an hexapod robot ($n = 12$, $m = 12$) to walk based on a measure of the overall speed. Both pendulum and robot are realised in the LpzRobots simulator [11].

For comparability, we follow the procedure in Ref. [5] and compare the performance to an RL controller configured as an actor-critic, see Eq. 1 for the actor output, and Eq. 3 for the learning rule. The critic is approximated by the relation $\dot{V}(t) \cong (V(t) - V(t - \triangle t))/\triangle t$ using the backward Euler approximation, which rises from the error signal (4)

$$\delta(t) = r(t) + \frac{1}{\triangle t}\left[\left(1 - \frac{\triangle t}{\tau}\right)V(t) - V(t - \triangle t)\right]. \tag{11}$$

The update of the w_i follows a gradient descent with respect to δ.

$$\dot{w}_i = \eta^C \delta(t)\frac{\partial V(x(t - \triangle t); w)}{\partial w_i}, \tag{12}$$

where η^C is a learning rate.

Actor and critic functions are implemented as a normalised Gaussian network. The sigmoid function is defined as $s(x) = \frac{2}{\pi}\arctan\left(\frac{\pi}{2}x\right)$. In the classic RL approach we use coloured noise with a correlation length of 0.1 as probing input with strength σ, in the case of self-exploring RL controller the same value σ is used to weigh the output $\mathbf{n}$ of the SO controller. The strength of the probing signal is weighted by σ, following the idea of [7], while the reward become bigger the probing input should become weaker, the value is calculated by $\sigma = \sigma_0 \min\left\{1, \max\left\{0, \frac{V_1 - V(t)}{V_1 - V_0}\right\}\right\}$ where V_0 and V_1 are the minimal and maximal levels of the reward. For the SO controller the activation function $g(\cdot) = \tanh(\cdot)$ is used.

4.1 Performance in a Toy Example

In a pendulum swing-up task, we use the same configuration as in Ref. [5] where the actor and the critic function are implemented in a 15×15 grid with the angle $\theta \in [-\pi, \pi]$ against the vertical line and the angular velocity $\omega \in [-2\pi, 2\pi]$. The reward function $r(\theta) = \cos(\theta)$ assumes the maximum at the upright and the minimum at the downward position of the pendulum. Each trial lasts for 20 seconds if $|\theta| < 5\pi$, otherwise a minimal reward is given for one second and the trial is reinitialised in a random state. The performance of the trial is measured by the time when the pendulum is in the range $|\theta| < \pi/4$. For the SO controller mostly the same setup is used. This problem is not trivial given the maximum applicable force to the pendulum $u^{\max} < m\gamma l$, this maximum force is multiplied to Eq. 5 with m as the mass of the pendulum, γ as the gravitational constant and l as the length from the pivot to the end of the pendulum, with a small enough $u^{\max}$ the pendulum has to build up momentum to be able to reach the upper position.

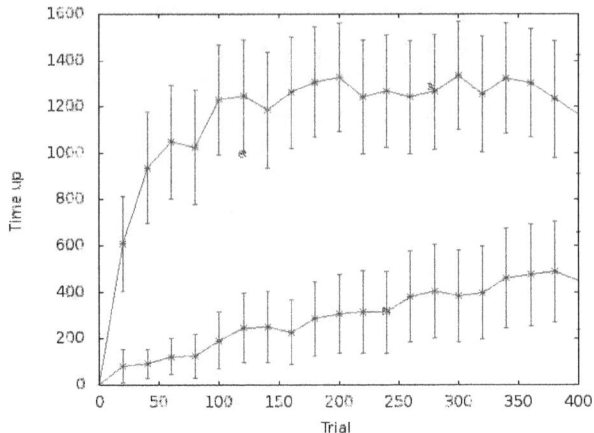

Fig. 2. Average of 50 experiments, each with 400 trials of the swing-up task with RL (upper trace) and with the combined controller (lower trace). The physical parameters are $m = 1$, $\gamma = 9.8$, friction $\mu = 0.01$, $u^{\max} = 5.0$. The parameters for the RL controller are $\tau = 1.0$, $\sigma_0 = 0.5$, $V_0 = -1$, $V_1 = 1$, $\triangle t = 0.02$. The learning rate for the C values were $\epsilon_c = 0.01$. The errorbars indicate deviations over 100 runs for each control task.

Fig. 2 shows the result of the average balance time per trial for 100 experiments with 400 trials each. Results discusion in section 5.

4.2 Self-organisation of Walking in an Hexapod

The hexapod (Fig. 3) resembles a insect with the three pairs of legs, two antennae and a thorax. A two-axis joint is placed where the legs meet the thorax allowing vertical and horizontal rotations, a servo motor actuate over each axis of the joint. In every axis a sensor measures angle θ with respect to the initial position and another sensor measures the angular velocity ω of the leg with pivot on the joint with the thorax. The joint between the femur and the tibia rotate in one axis with a damping action as springs.

The task of the hexapod robot was to improve its overall speed. A reward function is directly proportional to the speed in the (x,y)-plane. Due to the symmetry of the robot no particular movement direction is implied, i.e. in some trials the robot moved in direction of the antennae and in other ones in the opposite direction. The setup of the experiment is similar to the Pendulum, where a trial of 20 seconds is conducted by the controller, after that time the position and the velocities of each leg are set randomly. The performance of each trial is measured by the average speed of the trial. Again, normalised Gaussian networks are used as basis functions for each axis of the thorax-femur joints with 15×15 centers in the range $\left[-\frac{\pi}{8}, \frac{\pi}{8}\right] \times \left[-\frac{5}{4}\pi, \frac{5}{4}\pi\right]$ for vertical movements and $\left[-\frac{\pi}{4}, \frac{\pi}{4}\right] \times \left[-\frac{5}{4}\pi, \frac{5}{4}\pi\right]$ for horizontal movements of the legs. This high-dimensional task was performed over 4000 trials with SO controller and with noise probing signal, see Fig. 4.

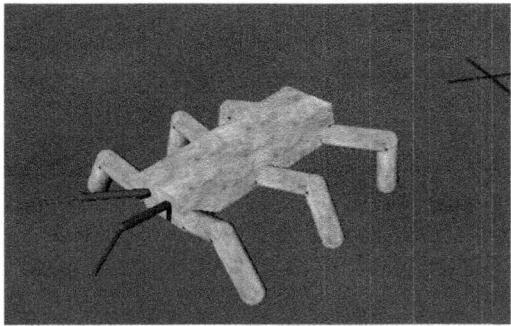

Fig. 3. Hexapod robot realised in the LpzRobots simulator. Joints between legs and thorax have two degrees of freedom and a servo motor for each axis. The joint in the middle of each leg contains only an unactuated spring.

5   Results and Discussion

5.1   Pendulum Results

A comparison of the results for a standard RL controller and for the SO controller are shown in Fig. 2. The swing-up task appears to be easily learned by the RL controller, its slope is steepest, a stable performance is reached earlier, and the total time spent in the upright position is longer. Interestingly, the SO controller never reaches a higher count of maximally rewarded states. This and the evidence that the behaviour is learned, shows that this controller continues to explore new states even if the maximum of the reward function has been already discovered. Because learning is driven merely by the correlation between exploratory action and utility function consistency, the results for this low-dimensional problem are little impressive, whereas in high dimensional tasks, where exploration is a less trivial problem, this SO controller will allow the robot to keep exploring such that local maxima of the expected reward or regions and directions with low gradients can be avoided more easily.

5.2   Hexapod Results

The results for the hexapod with the RL and the SO controller are shown in Fig. 4. As expected, the SO controller shows an advantage in the sense that throughout the experiment the increase of the average speed is higher than the maximal speed that was achieved by the RL controlled robot. It is, moreover, obvious that a stable performance is not clearly achieved by either approach, even after 4000 trials. The relatively high speed is produced by the SO controller even in spite the exploratory tendency of the SO controller, which can be considered as a further advantage of the explorative strategy.

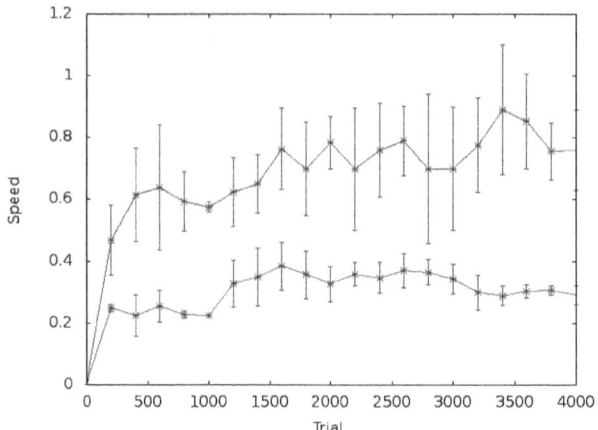

Fig. 4. Average speed for the hexapod with RL controller (lower trace) and with the combined controller (upper trace) during 4000 trials at a learning rate $\epsilon_C = 0.1$. The errorbars indicate deviations over three runs for each control task.

6 Conclusion

We have presented an integration of two approaches to the unsupervised generation of behaviour in robots. The interaction is based on an objective function that maximises the sensitivity of the learning systems with respect to mismatches in the utility function while simultaneously a RL component aims to maximise the future reward. We have tested our approach with two exemplary tasks of different complexity and have shown that

- the exploration induced by the SO controller may counteract the reward maximisation in an optimally tuned low dimensional task, while
- the SO controller seems to aid the learning process by guiding the exploration in a high dimensional task, and that
- the variable coherency of the action modulation in the SO controller improves the capability of the algorithm to escape local minima and flat regions of the goal function.

For comparability of the two variants of learning we ran the experiments with restarts after each trial. This is necessary in RL with random exploration but it is not required in the self-organized variant. If a stable performance is reached at any local or global optimum the sensitivity of the SO controller increases until the state of the systems escapes from the stationary behaviour. Restarting may not be an option for an autonomous robot such that an SO controller may be required here also for practical reasons.

Current work includes the comparison of the quality and frequency of the visited states as well as more systematic assessment of the scaling properties which are promising as we have shown recently in the context of guided self-organisation [9].

Acknowledgement. The design of the hexapod robot has been developed by Guillaume De Chambrier based on the data in Ref. [6]. We are very grateful to Georg Martius for helpful explanations of details of the LpzRobots simulator and to Frank Hesse for comments on the study. We want to thank Friedhelm Schwenker for the perfect organisation of the PSL 2011 workshop where the present paper has been first presented. This research has been funded by The Advanced Human Capital Program of the National Commission for Scientific and Technological Research (CONICYT) of the Republic of Chile.

References

1. Barto, A.G., Sutton, R.S., Anderson, C.W.: Neuronlike adaptive elements that can solve difficult learning control problems. IEEE Transactions on Systems, Man and Cybernetics 13, 834–846 (1983)
2. Der, R.: Self-organized acquisition of situated behavior. Theory Biosci. 120, 179–187 (2001)
3. Der, R., Michael Herrmann, J., Liebscher, R.: Homeokinetic approach to autonomous learning in mobile robots. VDI-Berichte, vol. 1679, pp. 301–306 (2002)
4. Der, R., Liebscher, R.: True autonomy from self-organized adaptivity. In: Workshop Biologically Inspired Robotics, Bristol (2002)
5. Doya, K.: Reinforcement learning in continuous time and space. Neural Computation 12, 219–245 (2000)
6. Ekeberg, Ö., Blümel, M., Büschges, A.: Dynamic simulation of insect walking. Arthropod Structure & Development 33, 287–300 (2004)
7. Gullapalli, V.: A stochastic reinforcement learning algorithm for learning real-valued functions. Neural Networks 3, 671–692 (1990)
8. Martius, G.: Goal-Oriented Control of Self-Organizing Behavior in Autonomous Robots. PhD thesis, Göttingen University (2010)
9. Martius, G., Herrmann, J.M.: Tipping the scales: Guidance and intrinsically motivated behavior. In: Proc. of Europ. Conf. on Artificial Life (2011)
10. Martius, G., Herrmann, J.M., Der, R.: Guided Self-Organisation for Autonomous Robot Development. In: Almeida e Costa, F., Rocha, L.M., Costa, E., Harvey, I., Coutinho, A. (eds.) ECAL 2007. LNCS (LNAI), vol. 4648, pp. 766–775. Springer, Heidelberg (2007)
11. Martius, G., Hesse, F., Güttler, F., Der, R.: Lpzrobots: A free and powerful robot simulator (2011), `robot.informatik.uni-leipzig.de`
12. Sutton, R.S.: Learning to predict by the methods of temporal differences. Machine Learning 3, 9–44 (1988)
13. Sutton, R.S., Barto, A.G.: Reinforcement learning: An introduction. MIT Press, Cambridge (1998); A Bradford Book
14. Wiener, N.: Cybernetics or Control and Communication in the Animal and the Machine. Hermann, Paris (1948)

Iterative Refinement of HMM and HCRF
for Sequence Classification

Yann Soullard and Thierry Artieres

Pierre and Marie Curie University, LIP6,
4 Place Jussieu, 75005 Paris, France
{yann.soullard,thierry.artieres}@lip6.fr

Abstract. We propose a strategy for semi-supervised learning of Hidden-state Conditional Random Fields (HCRF) for signal classification. It builds on simple procedures for semi-supervised learning of Hidden Markov Models (HMM) and on strategies for learning a HCRF from a trained HMM system. The algorithm learns a generative system based on Hidden Markov models and a discriminative one based on HCRFs where each model is refined by the other in an iterative framework.

1 Introduction

Sequence and signal labelling is a fundamental task in many application domains such as signal and speech recognition. We focus on sequence classification where one wants to assign a single label to an input sequence and aims at designing accurate semi-supervised learning (SSL) algorithms for discriminative markovian models such as Hidden Conditional Random Fields (HCRFs) [7]. Dealing with signal data requires using latent variable models, alike in Hidden Markov Models (HMMs). HMMs are the reference technology for dealing with sequences, while HCRFs may be viewed as a discriminative counterpart of HMMs, they are an extension of CRFs for dealing with hidden states [7].

Generative approaches (e.g. HMMs) rely on the learning of one model per class and build a distribution over observations. They may exhibit a higher bias and a lower variance than discriminative models [2] and they also may outperform discriminative models with small training datasets. Besides, generative models allow simple semi-supervised learning through the use of mixture models and of an EM learning scheme [5]. On the other hand, the discriminative approach (e.g. HCRFs) directly model the conditional probability distribution which is more related to the classification goal. They usually exhibit better asymptotic performance, when the training set size increases towards infinity, than generative models but the convergence to their optimal behaviour may be slower, meaning that generative models may reach their asymptotic performance faster than discriminative ones, i.e. with a smaller training set size [6]. Besides, semi-supervised training in discriminative models is less straightforward.

F. Schwenker and E. Trentin (Eds.): PSL 2011, LNAI 7081, pp. 92–95, 2012.

In particular designing SSL algorithms of HCRF is a difficult task. Although some works have been done on the more general task of semi-supervised learning of structured output predictors [3] and on the semi-supervised learning of Condition Random Fields (CRFs) [8], few works have concerned the semi-supervised training of Hidden-state CRFs (HCRFs). This situation motivated us to investigate a co-training like algorithm [1] where a generative system (HMMs) and a discriminative one (HCRFs) are iteratively refined. The idea is to rely on the relative easiness of SSL for HMMs and on recent initialization schemes of a HCRF from a HMM [4], [9]. In this algorithm HMMs make explicit use of unlabeled data while HCRF exploit them indirectly through the influence of a HMM trained in a semi-supervised setting.

2   Iterative Refinement of HMM and HCRF

A number of works have proposed methods for semi-supervised learning of HMMs or HCRFs [10]. For semi-supervised learning of HMMs for classification, a simple idea consists in learning a mixture model with one HMM component per class with labeled and unlabeled data using the EM algorithm. Semi-supervised training of HCRF is less simple though some approaches that have been proposed in the past for general structured output prediction models [3]. To overcome the difficulty of using unlabeled data in HCRF, we suggest to learn it using only labeled data while we exploit unlabeled data through the contribution of a HMM based system. To do that, we initialize the HCRF model from a HMM as proposed in [4] (we denote this strategy *HCRF Init*) or we used an alternative hybrid HMM-HCRF approach we proposed in [9] (we denote this as *HCRF Hybrid*). Whatever the method, the obtained HCRF indirectly depends on the unlabeled data through its initialization by a HMM trained in a semi-supervised manner.

Thus, we propose to learn iteratively two systems, a HMM system, and a HCRF system where each of them influence the other. The algorithm is illustrated in Figure 1. Initialization starts by SSL of HMMs which are then used to initialize HCRFs which are trained in a fully supervised way, as discussed above. Then, the algorithm iterates the two following steps : In a first step, the current HCRFs are used to label all unlabeled data and the HMMs are learned in a supervised way using all labeled and unlabelled data. Then, the HCRFs are learned on the labeled data, based on the HMM solution, either through initialization or with our hybrid modeling.

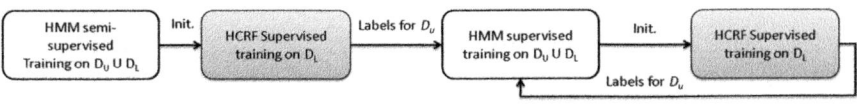

Fig. 1. Semi-supervised strategy embedding HMM and HCRF learning, D_L and D_U denote the sets of labeled and unlabeled training sequences

This strategy has some similarity with co-training method which was proposed by [1], where two classifiers are learned on two descriptions of the data and their predictions on unlabeled data are used to augment the training dataset of the other classifier. Provided that semi-supervised training is efficient for HMM and that HCRF outperform a HMM it is initialized from, such a strategy may be expected to work well.

3   Experiments

We performed experiments on chart pattern classification. A chart pattern is a particular shape of a stock exchange series of interest for financial operators. We used two databases, the first one (*CP4*) includes 448 samples corresponding to the 4 most popular patterns (*Head and Shoulders, Double Top, ...*). The second dataset *CP8* is a superset of CP4 with 4 more classes, it includes 892 patterns from 8 classes. Learning parameters are selected based on best results on the validation set and performance is measured on the test set. We use 10 labelled samples per class and 50 samples per class that we considered unlabeled. We report cross-validation results on 10 folds. HMM and HCRF of a class are left-right HMMs with the same topology. The pdf of a HMM state uses one diagonal covariance matrix Gaussian.

Figure 2 shows the evolution of performance of the HMM system and of the HCRF system as a function of the iteration number on the CP4 dataset. As may be seen, the final HCRF system significantly outperforms the semi-supervised HMM and the HCRF learned from it (i.e. 1st iteration result). More importantly, iterating the process allows building more discriminative HCRF and also more discriminative HMMs. The optimal number of iteration may be determined using a validation dataset. Then, we compare in Table 1 comparative results of supervised HMMs, supervised HCRFs initialized as in [4](HCRF *Init HMM*), supervised HCRFs trained with our hybrid algorithm in [9] (HCRF *Hybrid HMM*) and of the semi-supervised versions of these models learned with the algorithm in Figure 1. As may be seen semi-supervised training algorithm allows reaching higher recognition rates both for HCRF and HMMs.

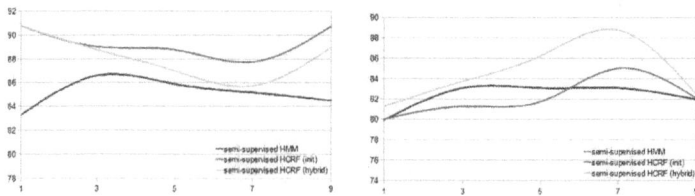

Fig. 2. Evolution of the performance of the HMM and of the HCRF system with iteration number on the training set (left) and on the test set (right)

Table 1. Comparison of diagonal covariance gaussian HMM, HCRF initialized based on learned HMMs and hybrid HCRF HMM model, and variants exploiting our semi-supervised learning strategy

Model	CP4	CP8
supervised HMM	80.0%	62.6%
supervised HCRF (*Init HMM*)	81.6%	63.75%
supervised HCRF (*Hybrid HMM*)	80.5%	64%
semi-supervised HMM	83.1%	64.4%
semi-supervised HCRF *Init*	85.0%	**67.5%**
semi-supervised HCRF *Hybrid*	**88.75%**	64.6%

4   Conclusion

We presented an iterative algorithm for learning accurately a Hidden-state CRF exploiting unlabeled data and a semi-supervised HMM system. Experimental results show that this strategy allows efficiently taking into account unlabeled data in HCRF training. A byproduct is that the HMM system may also benefit from this simultaneous training with a discriminative model.

References

1. Blum, A., Mitchell, T.: Combining labeled and unlabeled data with co-training. In: Conference on Computational Learning Theory. Morgan Kaufmann Pub. (1998)
2. Bouchard, G.: Bias-Variance Tradeoff in Hybrid Generative-Discriminative Models. In: International Conference on Machine Learning and Applications (2007)
3. Brefeld, U., Scheffer, T.: Semi-supervised learning for structured output variables. In: Proceedings of the 23th International Conference on Machine Learning, ICML 2006 (2006)
4. Gunawardana, A., Mahajan, M., Acero, A., Platt, J.C.: Hidden Conditional Random Fields for Phone Classification. In: Interspeech (2005)
5. Nigam, K., McCallum, A., Thrun, S., Mitchell, T.: Text Classification from Labeled and Unlabeled Documents using EM, pp. 103–134 (1999)
6. Ng, A.Y., Jordan, M.I.: On Discriminative vs. Generative Classifiers: A comparison of logistic regression and naive Bayes. In: NIPS (2001)
7. Quattoni, A., Wang, S., Morency, L.P., Collins, M., Darrell, T.: Hidden-state Conditional Random Fields. IEEE Transactions on Pattern Analysis and Machine Intelligence 29(10), 1848–1852 (2007)
8. Suzuki, J., Fujino, A., Isozaki, H.: Semi-supervised structured output learning based on a hybrid generative and discriminative approach. In: Proceedings of the 45th Annual Meeting of the Association for Computational Linguistics, ACL (2007)
9. Soullard, Y., Artieres, T.: Hybrid HMM and HCRF model for sequence classification. In: Proceedings of the European Symposium on Artificial Neural Networks, ESANN 2011 (2011)
10. Yu, D., Varadarajan, B., Deng, L., Acero, A.: Active learning and semi-supervised learning for speech recognition: A unified framework using the global entropy reduction maximization criterion. In: Computer Speech and Language (2009)

On the Utility of Partially Labeled Data for Classification of Microarray Data

Ludwig Lausser*, Florian Schmid*, and Hans A. Kestler**

Research Group Bioinformatics and Systems Biology
Institute of Neural Information Processing, University of Ulm, Germany
{ludwig.lausser,florian-1.schmid,hans.kestler}@uni-ulm.de

Abstract. Microarrays are standard tools for measuring thousands of gene expression levels simultaneously. They are frequently used in the classification process of tumor tissues. In this setting a collected set of samples often consists only of a few dozen data points. Common approaches for classifying such data are supervised. They exclusively use categorized data for training a classification model. Restricted to a small number of samples, these algorithms are affected by overfitting and often lack a good generalization performance. An implicit assumption of supervised methods is that only labeled training samples exist. This assumption does not always hold. In medical studies often additional unlabeled samples are available that can not be categorized for some time (i.e., "early relapse" vs. "late relapse"). Alternative classification approaches, such as semi-supervised or transductive algorithms, are able to utilize this partially labeled data. Here, we empirically investigate five semi-supervised and transductive algorithms as "early prediction tools" for incompletely labeled datasets of high dimensionality and low cardinality. Our experimental setup consists of cross-validation experiments under varying ratios of labeled to unlabeled examples. Most interestingly, the best cross-validation performance is not always achieved for completely labeled data, but rather for partially labeled datasets indicating the strong influence of label information on the classification process, even in the linearly separable case.

1 Introduction

In modern clinical studies the progress of a disease is monitored by DNA microarrays. These tools are high-throughput molecular biology devices for measuring thousands of gene expression levels simultaneously. The data collected within a clinical study usually does not exceed a few dozen gene expression profiles. These profiles can for example be used to discriminate the patients into clinical relevant groups (e.g. "inflammation" vs. "tumor"). In this setting a classifier performing this task has to deal with data of high dimensionality and low cardinality.

The standard learning scheme for training such a classifier is the supervised one. Here, a classifier is trained on a set of labeled samples. An implicit assumption of this scheme is that a training set of appropriate size exists.

---

* Contributed equally.
** Corresponding author.

F. Schwenker and E. Trentin (Eds.): PSL 2011, LNAI 7081, pp. 96–109, 2012.
© Springer-Verlag Berlin Heidelberg 2012

Clinical relevant classification tasks do not always perfectly fit into this basic supervised scenario. In many cases the unlabeled data is available years before the corresponding diagnoses. For example, it could be of interest how a patient reacts to a certain treatment. It is important to know if a patient will suffer from an "early relapse" or have a "late relapse" of a disease. Applying the standard supervised scheme, the earliest moment to start the analysis of the collected dataset is after receiving the last label. Often it is desirable to receive preliminary predictions within an earlier stage.

Alternative learning schemes, like semi-supervised learning, are able to handle partially labeled datasets. They utilize the positional information of a data point during training. Although semi-supervised algorithms fit better into the setting described above, they are normally applied in fields with much more available observations. So far it is unclear how these algorithms perform on small sample sizes.

In our study we investigate the usability of semi-supervised algorithms as early predictors for small (microarray) datasets. Five of these classifiers were tested on seven public available microarray datasets under varying conditions. We utilize an experimental setup consisting of adapted $l \times k$ cross-validation experiments allowing to assess the performance of semi-supervised and transductive algorithms under varying ratios of labeled to unlabeled examples.

2   Methods

In general a classifier c can be seen as a function mapping $c : \mathscr{X} \rightarrow \mathscr{Y}$ from an input space $\mathscr{X}$ to the space of class labels $\mathscr{Y}$. In the following only binary classifiers will be considered and the space of class labels will be fixed to the Boolean space $\mathscr{Y} := \{0,1\}$. Normally it is assumed that $\mathscr{X} \times \mathscr{Y}$ is associated with a fixed but unknown probability distribution. A common objective for a classifier is to minimize its *generalization risk* according to this distribution

$$\mathscr{R} = Pr(c(X) \neq Y). \tag{1}$$

Here (X,Y) denotes a random example drawn *iid* from $\mathscr{X} \times \mathscr{Y}$. As the distribution of $\mathscr{X} \times \mathscr{Y}$ is unknown, the generalization risk of this classifier has to be estimated according to a finite set $\mathscr{S}_{te} = \{(x'_i, y'_i)\}_{i=1}^{m'}$ of test samples.

$$R_{emp} = \frac{1}{m'} \sum_{(x',y') \in \mathscr{S}_{te}} \mathbb{I}_{[c(x') \neq y']}. \tag{2}$$

Here $\mathbb{I}_{[]}$ denotes the indicator function, which is equal to 1, if the condition in $[]$ is fulfilled and 0 otherwise. R_{emp} is called the *empirical risk*.

During an initial training phase a classifier has to be adapted to the current classification task. This is done according to a finite set of training examples $\mathscr{S}_{tr} = \{(x_i, y_i)\}_{i=1}^{m}$ with $\mathscr{S}_{tr} \cap \mathscr{S}_{te} = \emptyset$. Different learning paradigms exist, varying in how the available samples are incorporated. We will use $\mathscr{X}_{tr} := \{x'_i\}_{i=1}^{m}$ and $\mathscr{X}_{te} := \{x'_i\}_{i=1}^{m'}$ as additional notation to denote the unlabeled training and test samples.

2.1   Supervised Learning

Supervised learning schemes only incorporate knowledge from labeled samples. A prediction of a supervised classifier will be denoted by $c_{\mathscr{S}_{tr}}(x)$. They can be distinguished by how they use the training data in categorization.

Inductive learning: In this scheme the classifier c is chosen from a concept class $\mathscr{C}$ and adapted according to $\mathscr{S}_{tr}$ within a learning procedure l. Once trained, the classifier can be abstracted from the original training data; it can be independently applied on $\mathscr{S}_{te}$.

$$l(\mathscr{C}, \mathscr{S}_{tr}) \to c \tag{3}$$

Model-free learning: Training and application of a classifier can not be separated in this setting. The label of a single test sample x' is directly predicted according to measurements on $\mathscr{S}_{tr}$.

$$\mathscr{S}_{tr} \times x' \to \hat{y}' \tag{4}$$

2.2   Semi-supervised Learning

The term semi-supervised learning will here be used for algorithms incorporating knowledge from both labeled and unlabeled samples during their training phase. A prediction of such a classifier will be denoted by $c_{\mathscr{S}_{tr}, \mathscr{X}_{te}}(x)$. This category will subsume the real semi-supervised algorithms and the transductive learning algorithms.

Semi-supervised learning: This term is normally used for inductive algorithms that can also incorporate knowledge of unlabeled samples within their training. The final classifier is again independent of the training data used to adapt it and can be applied without knowing it.

$$l(\mathscr{C}, \mathscr{S}_{tr}, \mathscr{X}_{te}) \to c \tag{5}$$

Transductive learning: This can be seen as the generalization of model-free learning. Here, the label of a single test sample x' is determined according to measurements on the labeled and unlabeled training data.

$$\mathscr{S}_{tr} \times \mathscr{X}_{te} \to \hat{y} \tag{6}$$

2.3 $l \times k$ Cross-Validation

In supervised classification one standard evaluation method for datasets of small sample size is the k-fold cross-validation experiment (see e.g. [3, 7, 9]). The benefit of this method is its guarantee that each sample is used as training as well as test sample. Subsampling effects resulting in misleading performance measures are minimized.

For this experiment the available data of n samples is split into k-folds ($2 \leq k \leq n$) of approximately equal size (Figure 1). A subset of $k - 1$ folds is used as a labeled training set. The remaining fold is used as an independent test set. The procedure is repeated for each of the k possible splits into training sets $\mathscr{S}_{tr}^i$ and test sets $\mathscr{S}_{te}^i$, with $i \in \{1,\ldots,k\}$. In this way each sample is used once as an test sample; the cross-validation results in one prediction per datapoint. These predictions are then used to estimate the risk of the classifier.

$$R_{CV} = \frac{1}{n} \sum_{i=1}^{k} \sum_{(x,y) \in \mathscr{S}_{te}^i} \mathbb{I}\left[c_{\mathscr{S}_{tr}^i}(x) \neq y \right] \tag{7}$$

The estimate can be affected by the particular choice of splits. In order to minimize their influence, the k-fold cross-validation is repeated on l different permutations (runs) of the original dataset. The risk of the classifier is then estimated by the average over the l cross-validation errors. The final experiment is called a $l \times k$ cross-validation.

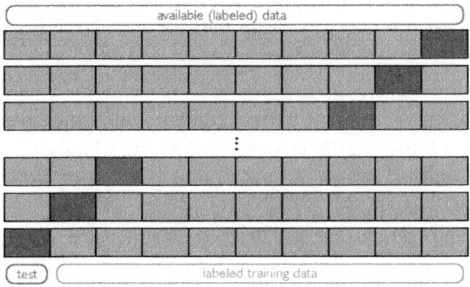

Fig. 1. Basic concept of a k-fold cross-validation. The available data is split into k folds of approximately equal size. The samples of $k-1$ folds are used as (labeled) training set for a classification model. The remaining fold is used as an independent set of test samples. The procedure is repeated for all k possible splits. The number of misclassifications over the whole dataset is used for estimating the risk of the classifier.

2.4   Cross-Validation Experiments for Semi-supervised Classifiers

In order to gain insight into the usability of semi-supervised algorithms as early prediction methods, two different cross-validation types were used (Figure 2).

Standard cross-validation: In this setting a classifier $c_{\mathscr{S}_{tr}^i, \mathscr{X}_{te}^i}(x)$ is adapted to all available samples. The labeled samples come from $\mathscr{S}_{tr}^i$ while the unlabeled samples come from $\mathscr{X}_{te}^i$. The tests are performed on $\mathscr{S}_{te}^i$.

$$R_{CV} = \frac{1}{n} \sum_{i=1}^{k} \sum_{(x,y) \in \mathscr{S}_{te}^i} \mathbb{I}\left[c_{\mathscr{S}_{tr}^i, \mathscr{X}_{te}^i}(x) \neq y \right] \tag{8}$$

Inverted cross-validation: The algorithms were also compared in a setting we call inverted cross-validation. Here for each fold i the $\mathscr{S}_{te}^i$ was used as labeled training set and the samples of $\mathscr{X}_{tr}^i$ were used unlabeled. The algorithm receives more unlabeled than labeled training examples. The error of a classifier is estimated according to

$$R_{CV-1} = \frac{1}{n(k-1)} \sum_{i=1}^{k} \sum_{(x,y) \in \mathscr{S}_{tr}^i} \mathbb{I}_{\left[c_{\mathscr{S}_{te}^i, \mathscr{X}_{tr}^i}(x) \neq y \right]}, \tag{9}$$

The results of the inverted cross-valdiation will be indexed by $l \times -k$. The learning task given by the inverted cross-validation setting better fits to the typical proportion of labeled and unlabeled training samples of known semi-supervised applications. Its benefit is the more systematic evaluation of the performance of a classifier than for example the evaluation done by experiments on randomly drawn splits.

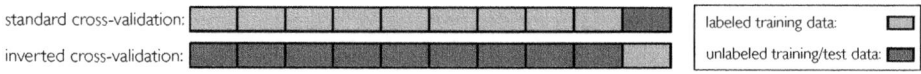

Fig. 2. Differences between the standard cross-validation and the inverted cross-validation: The figure shows the splits of the available data for the two kinds of experiments in the semi-supervised scenario. While the standard cross-validation setting utilizes $k-1$ folds as labeled training data and 1 fold as unlabeled training (test) data, the inverted cross-validation uses one fold as labeled training data and $k-1$ as unlabeled training (test) data.

2.5   Algorithms

We have tested following five algorithms on their usability as "early prediction tools":

Transductive support vector machines (tsvm) [10]: The algorithm applied here is a version of the standard (linear) inductive svm [16]. The basic strategy of both classification methods is to find a linear hyperplane ω maximizing the margin to the given samples. If it is not possible to separate the data correctly, a tradeoff between the misclassified datapoints (distance to margin) and the diminished margin has to be found.

In contrast to the inductive version, the class labels $\mathscr{Y}'$ of the test samples are directly included in the optimization process of the tsvm algorithm. They become estimated by solving an optimization task, described by the following system of linear equations:

$$\min_{\omega, \theta, \xi, \xi', \mathscr{Y}'} \quad \|\omega\|_2^2 + C \sum_{i=1}^{m} \xi_i + C' \sum_{i=1}^{m'} \xi_i'$$

$$\text{s.t.} \quad \forall_{i=1}^{m} : y_i(\omega^T x_i) - \theta \geq 1 - \xi_i, \xi_i \geq 0$$

$$\forall_{i=1}^{m'} : y_i'(\omega^T x_i') - \theta \geq 1 - \xi_i', \xi_i' \geq 0$$

The available labeled samples x_i and the unlabeled samples x'_i are separately treated within this optimization problem. For each kind of data there is a combination of cost parameter and distance measure, called C and ξ_i for the labeled samples and C' and ξ'_i for the unlabeled ones. The binary variables y'_i are chosen within the algorithm according to the solution of the optimization task. The cost parameters of this algorithm were fixed to a value of 1 in our experiments.

Penalized likelihood based pattern classification algorithm (plc) [2]: This algorithm estimates the likelihood $P_l = P(Y = 1|X = x_l)$ for each given sample. As the algorithm does not estimate a likelihood function, it belongs to the category of transductive algorithms. The estimates are determined in a penalized optimization task.

$$\min \quad J = \log(L) - \lambda S, \tag{10}$$

where L is the likelihood for the labeled samples $\mathscr{S}_{tr}$

$$L = \prod_{l=1}^{m} P_l^{y_l}(1 - P_l)^{1-y_l} \tag{11}$$

and S (smoothness) is a penalty on the roughness of the estimations

$$S = \frac{1}{K} \sum_{l=1}^{m+m'} \sum_{l' \in \mathbb{K}(x_l)} (P_l - P_{l'})^2. \tag{12}$$

The smoothness of each prediction is calculated according to the neighbourhood $\mathbb{K}$. The size K of this neighbourhood is determined within the algorithm. As proposed in [2], we set parameter $\lambda = 0.4$.

Transductive k-nearest neighbors classifier (tknn) [14]: This version of the k-nearest neighbor classifier (e.g. [8]) determines the label of a single sample according to measurements on its k_l labeled and k_u unlabeled neighbors. The influence of the single samples on the classification of a datapoint x_i is thereby regulated according to a weight vector w_i.

$$w_{ij} = \begin{cases} K(x_i, x_j), & \text{if } x_j \in \mathscr{X}_{tr} \wedge x_j \in \mathbb{K}_l(x_i, k_l) \\ aK(x_i, x_j), & \text{if } x_j \in \mathscr{X}_{te} \wedge x_j \in \mathbb{K}_u(x_i, k_u) \\ 0, & \text{otherwise} \end{cases} \tag{13}$$

Here $K(x_i, x_j)$ denotes following distance kernel function

$$K(x_i, x_j) = \frac{1}{\sqrt{2\pi}h} \exp\left(-\frac{||x_i - x_j||^2}{2h^2}\right). \tag{14}$$

The label of a sample x_i is determined within a label propagation process iteratively calculating the class membership probabilities p_{ir}, $r \in \{0, 1\}$.

$$p_{ir}^{[t+1]} \leftarrow \sum_{j=1}^{m+m'} v_{ij} p_{jr}^{[t]} \tag{15}$$

Here v_{ij} corresponds to the row normalized value of w_{ij}. The initial class membership probabilities is initialized by 0.5 for unlabeled samples and fixed to 0 and 1 for labeled ones. The propagation process is repeated until the last class membership probability of an unlabeled test sample has converged. We have fixed the number of labeled (unlabeled) neighbors to $k_l = 1$ ($k_u = 3$). The influence of the unlabeled neighbors was regularized by $a \in \{0.3, 0.7, 1.0\}$.

Yarowsky's algorithm (yar) [18]: This algorithm is a general iterative procedure for modifying an inductive classifier $c_{\mathscr{S}_{tr}}$ into a semi-supervised one. The inductive algorithm must therefore be able to give confidence values $p_{\mathscr{S}_{tr}}$ for its predictions. We used a svm, which returns class probabilities, as a base classifier [12] . Yarowsky's algorithm iteratively includes unlabeled samples into the (labeled) training set, if they allow a prediction above a fixed confidence level d.

$$\mathscr{S}_{tr}^{[t+1]} = \mathscr{S}_{tr}^{[0]} \cup \{(x', \hat{y}) \mid x' \in \mathscr{X}_{te}, \hat{y} = c_{\mathscr{S}_{tr}^{[t]}}(x'),\ p_{\mathscr{S}_{tr}^{[t]}}(x') \geq d\} \qquad (16)$$

The classifier is retrained on the modified training sets until $\mathscr{S}_{tr}^{[t+1]} = \mathscr{S}_{tr}^{[t]}$. In our experiments the confidence level is chosen from $d \in \{0.6, 0.7, 0.8, 0.9\}$.

Mincut algorithm (mc) [5]: This algorithm is based on a weighted graph connecting the samples of the dataset. The graph is extended by a node for each class label of the dataset. These nodes are connected with all samples of the corresponding class. The weights of these edges are set to infinity. During the training process the graph is pruned according to a max-flow algorithm. The remaining paths to one of the label nodes determine the labels of the samples. The graph we have chosen in our experiments is based on the dataset's distribution of pairwise (Euclidian) distances. An edge between two datapoints is drawn, if the corresponding distance is smaller than the q-quantile of this distribution ($q \in \{0.1, 0.2, 0.3\}$).

3 Experimental Setup

We compared the five semi-supervised algorithms mentioned before in a series of cross-validation experiments on seven microarray datasets (see Table 1). The series include $10 \times k$ cross-validations for k = 10,. . .,2 and inverted $10 \times -k$ cross-validations for $k = 2, \ldots, 10$. The single experiments differ in their fold number and the number of available training and test samples; while the number of (unlabeled) training samples decreases with k, the number of (labeled) test samples increases. An overview on the available positive and negative samples in the $10 \times k$ setting can be found in Figure 3. Over the seven datasets the mean number of labeled training samples per fold varies from 25.6 to 91.8 within the 10×10 experiment and 14.0 to 51.0 in the 10×2 experiment; the corresponding mean number of unlabeled test samples per fold vary from 2.8 to 10.2 (10×10) and 14.0 to 51.0 (10×2). In the inverted cross-validation the numbers of training and test samples are reversed.

Table 1. Key properties of the utilized datasets

Dataset	Features	Positive samples	Negative samples
Armstrong (*AR*) [1]	12582	24	48
Bittner (*BI*) [4]	8067	19	19
Nutt (*NU*) [11]	12625	14	14
Pancreas (*PA*) [6]	169	37	25
Shipp (*SH*) [13]	7129	58	19
Singh (*SI*) [15]	12600	52	50
West (*WE*) [17]	7129	25	24

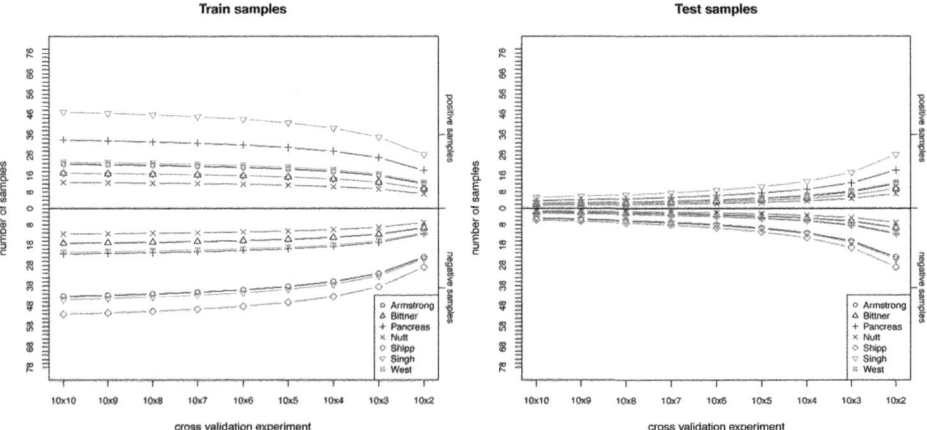

Fig. 3. Number of samples (microarray datasets): The figure shows the influence of the chosen k within a standard $l \times k$ cross-validation experiment on the number of available (labeled) training and (unlabeled) test examples per fold. While the number of (labeled) training examples increases with the number of folds, the number of (unlabeled) test examples decreases. The number of (labeled) training samples within a standard $l \times k$ cross-validation is corresponding to the number of (unlabeled) test samples within a inverted $l \times -k$ cross-validation and vice versa.

4 Results

The results of the cross-validation experiments can be found in Figures 4 and 5. The accuracies were additionally compared to the results of a "prevalence" classifier always predicting the class label of the larger class. The accuracy of the constant classifier can be seen as a lower bound for a beneficial ("meaningful") classification performance. In the case of imbalanced data, this bound is tighter than the 50% bound. For the semi-supervised classifiers following results could be observed:

Two of the algortihms, *tsvm* and *plc*, showed a better performance than the "prevalence" classifier on all datasets. Despite of its performance in the 10×-9 and 10×-10 cross-validation experiment on *SH*, the same is true for the *tknn* algorithm. For some datasets, the performance of *yar* and *mc* did not cross the minimal accuracy level; while

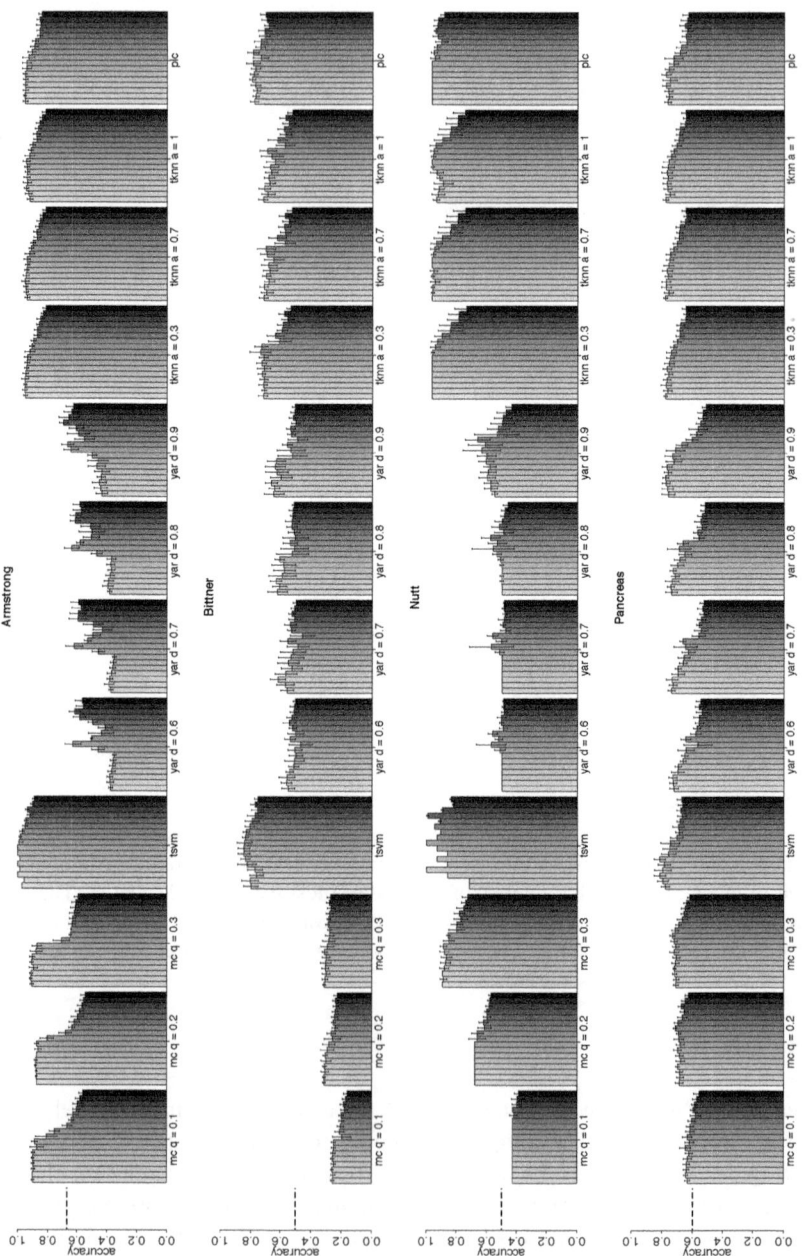

Fig. 4. Results of the $10 \times k$ cross-validation experiments ($k \in 10, \ldots, \pm 2, \ldots, -10$) for datasets Armstrong, Bittner, Nutt, Pancreas. A legend for the different experimental setups can be found in Figure 5.

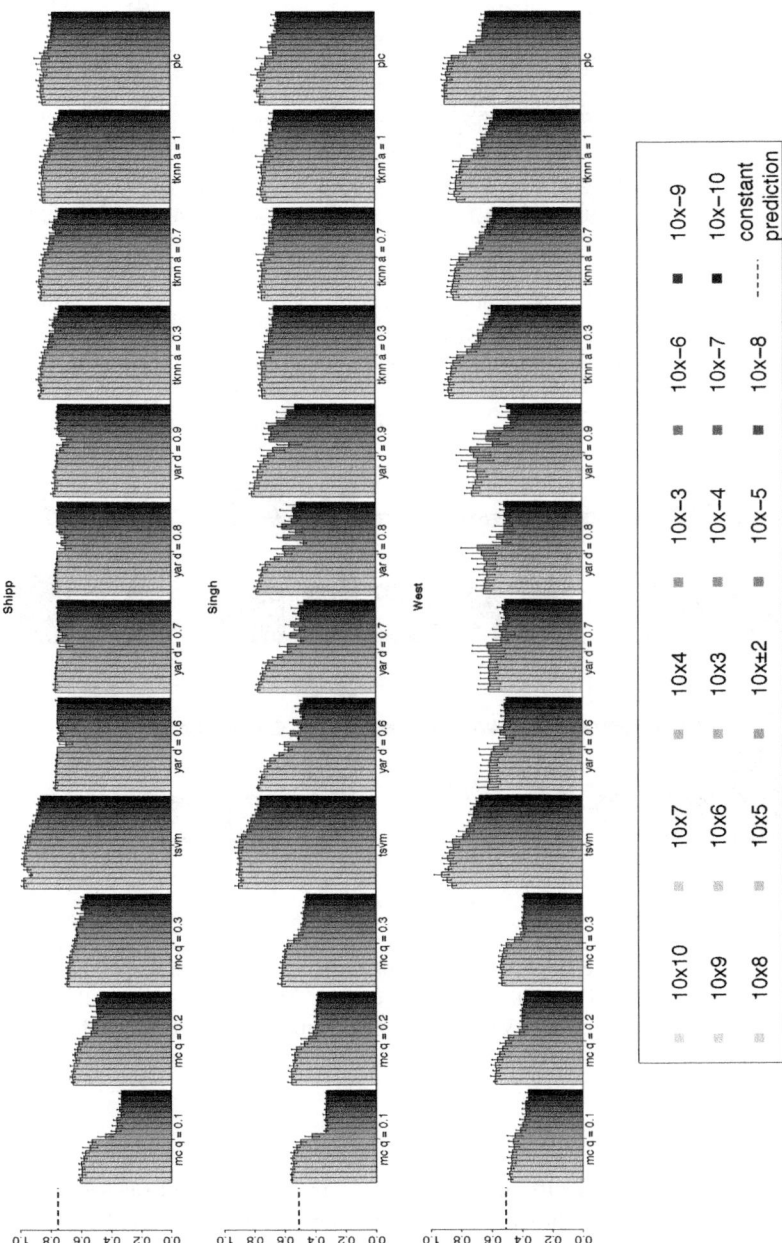

Fig. 5. Results of the $10 \times k$ cross-validation experiments ($k \in 10, \ldots, \pm 2, \ldots, -10$) for datasets Shipp, Singh, West

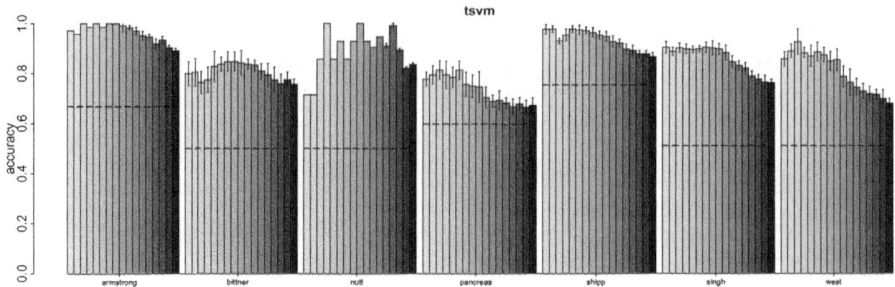

Fig. 6. Results of the $10 \times k$ cross-validation experiments ($k \in 10,\ldots,\pm 2,\ldots,-10$): The figure summarizes the results of the tsvm. A legend for the different experimental setups can be found in Figure 6. The dotted line corresponds to the results of a "prevalence" classifier.

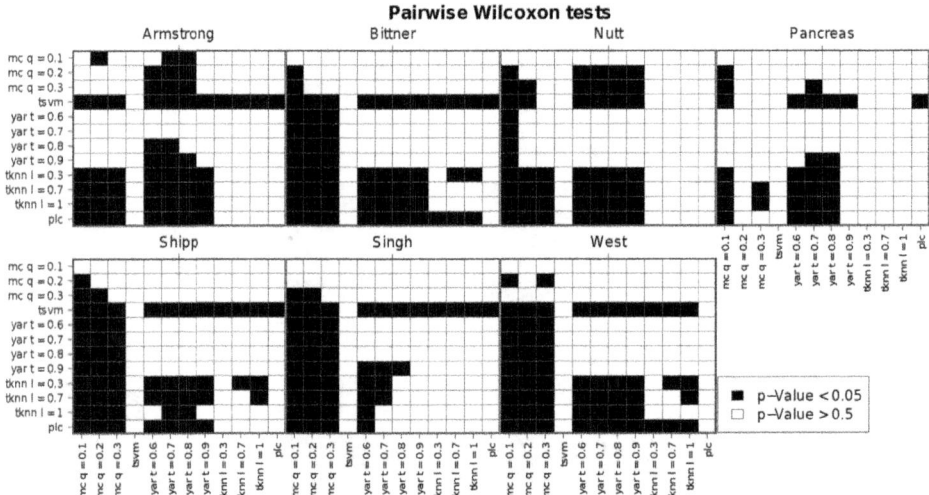

Fig. 7. Results of paired one-sided Wilcoxon rank tests done for the cross-validation accuracies (over all k) of every pair of algorithms on each dataset. The black color in cell ij denotes that the median cross-validation accuracy of algorithm i is significantly higher than the median cross-validation accuracy of algorithm j. For each dataset, the tests were corrected for multiple-testing (Bonferroni $n = 132$).

mc did not attain good results on *BI* and *SH*, *yar* did not excel on *AR*. On the other datasets these algorithms showed a low performance in the inverted cross-validation setting and again did not achieve the minimal accuracy level. In general lower accuracies were achieved within the inverted cross-validations than in the standard cross-validations. An exception is *yar* for the datasets *AR* and *NU*.

The *plc* is with a mean range (max accuracy − min accuracy) of about 13.5% the steadiest of the analysed algorithms. The mean range of *yar* (over all *t*) is with about

18.5% the largest. The other algorithms *tsvm*, *tknn* and *mc* achieved mean ranges of 16.2%, 17.0% (over all *l*), and 16.5%.

We applied paired one-sided Wilcoxon rank sum tests on the cross-validation accuracies (over all *k*) of every pair of algorithms on each dataset (Figure 7). The null hypothesis was that the first classifier has an equal or lower median cross-validation accuracy than the second. For each dataset, the tests were corrected for multiple-testing (Bonferroni $n = 132$). According to these tests, there was no algorithm which was significantly better than *tsvm*. The *tsvm* itself was significantly better than all other tested algorithms on four datasets (*AR, BI, SI, SH*). The *plc* was only outperformed by the *tsvm* on five datasets (*AR, BI, PA, SH, SI*). The *tknn* was not outperformed on the datasets *NU* and *PA*. *tknn* completely outperformed *yar* on the datasets *AR, BI, NU* and *WE*. The *mc* algorithm was outperformed by all algorithms on the datasets *BI, SH, WE* and *SI*.

5   Discussion

The major challenge in our settings is the low cardinality of the datasets ($\lesssim 100$) limiting the number of available labeled and unlabeled training samples. Although this is an unusual constraint for semi-supervised learning, some of the algorithms achieved good classification results in our experimental setting. The results on the standard cross-validation experiments were mostly better than those of the inverted ones. Coupled to a smaller number of labeled training samples, the results gained on the inverted cross-validation indicate that the lack of labeled training samples can often not be compensated by an increasing number of samples (which do not have a label). Nevertheless, and most interestingly the best cross-validation performance is not always achieved for completely labeled data, but rather for partially labeled datasets indicating the strong influence of label information on the classification process.

The lower performance of Yarowsky's algorithm and the mincut strategy can may be related to the small number of available samples. The iterative process of Yarowsky's algorithm is controlled by confidence predictions for the single data points. Related to the distance between the samples and the decision boundary these confidence predictions get less distinguishable and less informative in high dimensional settings.

The initial graph constructed by the mincut strategy was also effected by the small sample sizes. Here the majority of unlabeled test samples built separate subgraphs which were not connected to one of the label nodes. In this case the algorithm is not able to determine the class label of these samples.

The classifiers *tsvm* and *plc* showed better accuracies than a constant classifier throughout all experiments. The same holds true for *tknn* except for two single experiments. These three algorithms can therefore be used as early predictors. Finally, the *tsvm* achieved the best classification performance in our study followed by *plc* and *tknn*.

The accuracy of this algorithm is summarized in in Figure 6. Besides the overall trend of receiving higher accuracies for higher values of *k* an additional behavior of this algorithm can be seen. In some of the settings the best classification results is not directly achieved for $k = 10$. Better results can be found for slightly smaller values of *k*. We believe that this is not a direct effect of the tradeoff between labeled and unlabeled samples. Disturbances such as measurement and label noise seem to be related to this behavior with different severity. The benefit of this diminished label information even in the linearly separable case can serve as a starting point for future work.

Acknowledgments. This work was funded in part by a Karl-Steinbuch grant to FS, the German federal ministry of education and research (BMBF) within the framework of the program of medical genome research (PaCa-Net; project ID PKB-01GS08) and the framework GERONTOSYS (Forschungskern SyStaR, project ID 0315894A) to HAK. The responsibility for the content lies exclusively with the authors.

References

1. Armstrong, S.A., Staunton, J.E., Silverman, L.B., Pieters, R., den Boer, M.L., Minden, M.D., Sallan, S.E., Lander, E.S., Golub, T.R., Korsmeyer, S.J.: Mll translocations specify a distinct gene expression profile that distinguishes a unique leukemia. Nature Genetics 30(1), 41–47 (2002)
2. Atiya, A.F., Al-Ani, A.: A penalized likelihood based pattern classification algorithm. Pattern Recognition 42, 2684–2694 (2009)
3. Bishop, C.M.: Pattern Recognition and Machine Learning. Information Science and Statistics. Springer, Secaucus (2006)
4. Bittner, M., Meltzer, P., Chen, Y., Jiang, Y., Seftor, E., Hendrix, M., Radmacher, M., Simon, R., Yakhini, Z., Ben-Dor, A., Sampas, N., Dougherty, E., Wang, E., Marincola, F., Gooden, C., Lueders, J., Glatfelter, A., Pollock, P., Carpten, J., Gillanders, E., Leja, D., Dietrich, K., Beaudry, C., Berens, M., Alberts, D., Sondak, V.: Molecular classification of cutaneous malignant melanoma by gene expression profiling. Nature 406(6795), 536–540 (2000)
5. Blum, A., Chawla, S.: Learning from labeled and unlabeled data using graph mincuts. In: Brodley, C.E., Danyluk, A.P. (eds.) ICML 2001 Proceedings of the Eighteenth International Conference on Machine Learning, pp. 19–26. Morgan Kaufmann, San Francisco (2001)
6. Buchholz, M., Kestler, H.A., Bauer, A., Böck, W., Rau, B., Leder, G., Kratzer, W., Bommer, M., Scarpa, A., Schilling, M.K., Adler, G., Hoheisel, J.D., Gress, T.M.: Specialized DNA arrays for the differentiation of pancreatic tumors. Clinical Cancer Research 11(22), 8048–8054 (2005); HAK and MB contributed equally
7. Duda, R.O., Hart, P.E., Stork, D.G.: Pattern Classification, 2nd edn. Wiley, New York (2001)
8. Fix, E., Hodges Jr., J.L.: Discriminatory Analysis: Nonparametric Discrimination: Consistency Properties. Technical Report Project 21-49-004, Report Number 4, USAF School of Aviation Medicine, Randolf Field, Texas (1951)
9. Hastie, T., Tibshirani, R., Friedman, J.H.: The Elements of Statistical Learning, corrected edn. Springer, Heidelberg (2003)
10. Joachims, T.: Transductive Inference for Text Classification using Support Vector Machines. In: Bratko, I., Dzeroski, S. (eds.) Proceedings of ICML 1999, 16th International Conference on Machine Learning, pp. 200–209. Morgan Kaufmann Publishers, San Francisco (1999)
11. Nutt, C.L., Mani, D.R., Betensky, R.A., Tamayo, P., Cairncross, J.G., Ladd, C., Pohl, U., Hartmann, C., McLaughlin, M.E., Batchelor, T.T., Black, P.M., von Deimling, A., Pomeroy, S.L., Golub, T.R., Louis, D.N.: Gene expression-based classification of malignant gliomas correlates better with survival than histological classification. Cancer Research 63(7), 1602–1607 (2003)
12. Platt, J.: Probabilistic outputs for support vector machines and comparison to regularized likelihood methods. In: Bartlett, P.J., Schölkopf, B., Schuurmans, D., Smola, A.J. (eds.) Advances in Large Margin Classifiers. MIT Press (2000)

13. Shipp, M.A., Ross, K.N., Tamayo, P., Weng, A.P., Kutok, J.L., Aguiar, R.C.T., Gaasenbeek, M., Angelo, M., Reich, M., Pinkus, G.S., Ray, T.S., Koval, M.A., Last, K.W., Norton, A., Lister, T.A., Mesirov, J., Neuberg, D.S., Lander, E.S., Aster, J.C., Golub, T.R.: Diffuse large b-cell lymphoma outcome prediction by gene-expression profiling and supervised machine learning. Nature Medicine 8(1), 68–74 (2002)
14. Shu, L., Wu, J., Yu, L., Meng, W.: Kernel-Based Transductive Learning with Nearest Neighbors. In: Li, Q., Feng, L., Pei, J., Wang, S.X., Zhou, X., Zhu, Q.-M. (eds.) APWeb/WAIM 2009. LNCS, vol. 5446, pp. 345–356. Springer, Heidelberg (2009)
15. Singh, D., Febbo, P.G., Ross, K., Jackson, D.G., Manola, J., Ladd, C., Tamayo, P., Renshaw, A.A., D'Amico, A.V., Richie, J.P., Lander, E.S., Loda, M., Kantoff, P.W., Golub, T.R., Sellers, W.R.: Gene expression correlates of clinical prostate cancer behavior. Cancer Cell 1(2), 203–209 (2002)
16. Vapnik, V.N.: Statistical Learning Theory. Wiley, New York (1998)
17. West, M., Blanchette, C., Dressman, H., Huang, E., Ishida, S., Spang, R., Zuzan, H., Olson, J.A., Marks, J.R., Nevins, J.R.: Predicting the clinical status of human breast cancer by using gene expression profiles. Proceedings of the National Academy of Science of the United States of America 98(20), 11462–11467 (2001)
18. Yarowsky, D.: Unsupervised word sense disambiguation rivaling supervised methods. In: Uszkoreit, H. (ed.) ACL 1995 Proceedings of the 33rd Annual Meeting on Association for Computational Linguistics, pp. 189–196. Association for Computational Linguistics, Stroudsburg (1995)

Multi-instance Methods for Partially Supervised Image Segmentation

Andreas Müller and Sven Behnke

Autonomous Intelligent Systems
Department of Computer Science
University of Bonn
53113 Bonn, Germany
amueller@ais.uni-bonn.de, behnke@cs.uni-bonn.de

Abstract. In this paper, we propose a new partially supervised multi-class image segmentation algorithm. We focus on the multi-class, single-label setup, where each image is assigned one of multiple classes. We formulate the problem of image segmentation as a multi-instance task on a given set of overlapping candidate segments. Using these candidate segments, we solve the multi-instance, multi-class problem using multi-instance kernels with an SVM. This computationally advantageous approach, which requires only convex optimization, yields encouraging results on the challenging problem of partially supervised image segmentation.

1 Introduction

The task of multi-class image segmentation is to create a pixel-wise labeling of an input image into regions belonging to one of several semantic classes. Most algorithms for this setting work with strong supervision: a pixel-wise labeling of training images. Methods that are used in this setting include random forests [24] and support vector machines (SVM). Usually the output of such algorithms is further processed by a conditional random field (CRF [14, 11, 13]). While these methods reach high accuracy, it is very time consuming to create pixel-level ground truth for real-world applications. This problem can be addressed in several ways: the LabelMe effort [23] tries to use the "wisdom of crowds" to obtain human labelings. Another possibility is to use only weak supervision, which is the approach we follow here.

In the weakly supervised setting, the ground truth for a given image is a list of semantic classes that occur in this image, instead of a pixel-level labeling as in the strongly supervised setting. Image-level labels are much easier to obtain, e.g. through online image libraries such as flickr and facebook.

The task of multi-class segmentation is often split up in a segmentation and a recognition part. Random forest methods often classify each pixel separately and segment using predicted classes [24] while SVM-based methods often work on an over-segmentation of the image, called superpixels [14, 11]. Superpixels avoid the computational burden of classifying each pixel separately, but have two drawbacks:

F. Schwenker and E. Trentin (Eds.): PSL 2011, LNAI 7081, pp. 110–119, 2012.
© Springer-Verlag Berlin Heidelberg 2012

1. A single superpixel does not provide enough context for classification [9].
2. Segment boundaries are decided on the lowest level by generating superpixels. This decision cannot be corrected afterwards [12].

In our approach, we work with a set of candidate segments, generated using constrained parametric min-cuts [2]. For each image, these segments are a set of overlapping, object-like regions, which serve as candidates for object locations.

We formulate weakly supervised multi-class image segmentation as a multi-instance problem, based upon candidate segments. In multi-instance learning [6], each training example is given as a multi-set of instances, called a bag. Each instance is represented as a feature vector x and a label y. A bag is labeled positive if it contains at least one positive example, and negative otherwise. During training, only the labels of the training bags, not of the instances inside the bags, are known. The goal is to learn a classifier for unseen bags. Formally, let $\mathcal{X}$ be the set of instances. To simply notation, we assume that bags are simply sets, not multi-sets. Then a bag is an element of the power set $2^{\mathcal{X}}$ and the task is to learn a function

$$f_{MI}\colon 2^{\mathcal{X}} \to \{-1, +1\} \tag{1}$$

from a set of training examples of the form (X_i, y_i) with bags $X_i \subset \mathcal{X}$ and labels $y_i \in \{-1, +1\}$. The f_{MI} function stems from the so-called underlying concept, given by an (unknown) function $f_I\colon \mathcal{X} \to \{-1, +1\}$, with

$$f_{MI}(X) = \max_{x \in X} f_I(x). \tag{2}$$

Sometimes, the goal of finding f_{MI} is extended to finding labels not only on bag-level but also for all the instances within a bag [17, 31], i.e. finding f_I.

Even though finding f_I is sometimes included in the task statement, there has been very little work that actually reported accuracy on instance label prediction. Part of the reason for this might be that for many of the datasets used in multi-instance learning no ground truth exists.

We look explicitly at accuracy on instance-level since we are interested in actually segmenting images, not just classifying them. For multi-class image segmentation, there are some hand-labeled datasets that provide ground truth on pixel level. We use this ground truth to evaluate the performance of our method. This approach does not exactly correspond to instance-level ground truth – since the instances are segments, not pixels – but relates to it closely.

In this work, we explore the application of multi-instance learning algorithms to the task of partially supervised image segmentation. Multi-instance learning is a natural formulation for image classification and has been successfully applied in this task [35]. We propose to go a step further and apply multi-instance learning to the task of object-class segmentation in natural images. To our knowledge, all previous methods in the field use strong supervision, meaning manual pixel-wise annotation of training images. This approach does not scale to larger datasets, especially if one expects consistency and quality in the segmentations.

2   Related Work

2.1   Proposal Object Segments

Most work on multi-class segmentation focuses on strong supervision on super-pixel level. There is still little work on using candidate segments. The method we use for generating candidate segments is Constraint Parametric Min-Cuts (CPMC) from [2]. This method creates a wide variety of overlapping segments. Support vector regression (SVR) is trained on these segments to estimate the overlap of segments with ground truth object-class labeling from the Pascal VOC dataset [8]. This provides a ranking of candidate segments, according to how "object-like" they are, which allows for selecting only a limited number of very object-like segments. The method performed well on a variety of datasets. A similar   approach   was   investigated   by [7].

2.2   Multi-instance Methods

Multi-instance learning was formally introduced in [6]. Since then, many algorithms were proposed to solve the multi-instance learning problem using many different approaches [1, 10, 34, 18, 33, 21, 15, 4]. We will discuss only those that are relevant to this work.

[10] introduced the concept of a multi-instance kernel on bags, defined in terms of a kernel on instances. The basic principle of multi-instance kernel is similar to a soft-max over instances in each bag. This can be viewed as approximating the kernel value of the "closest pair" given by two bags. They show that the multi-instance kernel is able to separate bags if and only if the original kernel on instances is able to separate the underlying concepts. The method of [10] has a particular appeal in that it neatly transforms a multi-instance problem into a standard classification problem by changing the kernel. The downside of this approach is that it does not directly label instances, only bags.

[34] explicitly address non-i.i.d. labels, leading to an algorithm that can take advantage of correlations inside bags. Computational costs of their algorithm does not scale well with the number of instances, although a heuristic algorithm is proposed to overcome this restriction. [34] demonstrated only a slight advantage of their algorithm over the MI-kernel of [10], so we use the MI-kernel for better scalability.

[17] compute likelihood ratios for instances, giving a new convex formulation of the multi-instance problem. Using these likelihood ratios, classification can be performed directly on the instances, provided an appropriate threshold for classifying instances as positive is known. We circumvent this problem by applying the same classifier to instances and bags, thereby obtaining hard class decisions for each instance.

2.3   Semantic Scene Segmentation via Multi-instance Learning

Recently, several methods have been proposed to obtain semantic segmentations of images using only image-level supervision [29, 27, 28]. [29], for example, report impressive results on the MSRC dataset.

While semantic segmentation is closely related to multi-class image segmentation, there are important distinctions: In semantic segmentation, each pixel has a semantic annotation, also containing "background" classes like "sky", "grass" and "water". In multi-class image segmentation, the focus is on objects, with possibly large parts of the image being labeled as unspecific "background". The unspecific background class contains much more clutter than for example "grass" and is therefore much harder to model. This makes disseminating the interesting part in multi-class object recognition challenging, since it is not necessary possible to identify non-object regions easily.

3   Multi-instance Kernels for Image Segmentation

3.1   Constraint Parametric Min Cuts (CPMC)

To generate proposal segments, we use the CPCM framework from   [2]. We construct initial segments using graph cuts, on the image graph. The energy function for these cuts uses pixel color and the response of the global probability of boundary (gPb) detector [20]. As much as ten thousand initial segments are generated from foreground and background seeds. A fast rejection based on segment size and ratio cut [30] reduced these to about 2000 overlapping segments per image. Then, the segments are ranked according to a measure of object-likeness that is based on region and Gestalt properties. This ranking is computed using an SVR model [2], which is available online. For computing the global probability of boundary (gPb), we used the CUDA implementation of [3], instead of the original one, for speed.

3.2   Multi-instance Learning Using MI-Kernels

Since scalability is very important in real-world computer vision applications, and natural images might need hundreds of segments to account for all possible object boundaries, we use the efficient multi-instance kernels [10]. Multi-instance kernels are a form of set kernels that transform a kernel on instance level to a kernel on bag level. We reduce the multi-instance multi-class problem to a multi-instance problem by using the one-vs-all approach.

With k_I denoting a kernel on instances $x, x' \in \mathcal{X}$, we define the corresponding multi-instance kernel k_{MI} on bags $X, X' \in 2^{\mathcal{X}}$ as

$$k_{MI}(X, X') := \sum_{x \in X, x' \in X'} k^p(x, x'), \qquad (3)$$

where $p \in \mathbb{N}$ is a parameter [10]. As we use the RBF-kernel k_{rbf} as kernel on $\mathcal{X}$ and powers of RBF-kernels are again RBF-kernels, we will not consider p in the following.

We normalize the kernel k_{MI} [10] using

$$k(X, X') := \frac{k_{MI}(X, X')}{\sqrt{k_{MI}(X, X)k_{MI}(X', X')}}. \tag{4}$$

Training an SVM with this kernel produces a bag-level classifier for each class, which we will refer to as MIK. This procedure is very efficient since the resulting kernel matrix is of size number of bags, which is much smaller than a kernel matrix of size number of instances, as is commonly used in the literature [1, 22, 32]. Another advantage over other methods is, that it uses a single convex optimization, whereas other approaches often use iterative algorithms [1] or need to fit complex probabilistic models [31].

While using MIK has many advantages, it produces only an instance-level classifier. We propose to transform a bag-level classifier f_{MI} as given by the SVM and Equation (3) into an instance-level classifier by setting $f_I(x) := f_{MI}(\{x\})$, in other words, by considering each instance as its own bag.

3.3   Segment Features

To describe single segments, we make use of densely computed SIFT [19] and ColorSIFT [26] features, from which we compute bag of visual word histograms. Additionally, we use histograms of oriented gradients [5] on the segments. We use RBF-kernels for all of the features, constructing one MI-kernel per feature. These are then combined using multiple kernel learning to produce a single kernel matrix. This kernel matrix can then be used for all classes, making classification particularly efficient.

3.4   Combining Segments

The framework described above yields an image-level and a segment-level classi-fication. In our setup, each segment might be given multiple labels. To obtain a pixel-level object-class segmentation, we have to combine these. When building the segmentation for a given image, we only consider classes whose presence was predicted on image level. Since we do not make use of the ground truth segmen-tation during training, we cannot learn an optimal combination as in [16] but perform a simple majority vote instead. We merge segments into pixel-level class labels by setting the label y_x of a pixel x according to

$$y_x = \text{argmax}_{y \in Y} \#\{S_i | p \in S_i \wedge y_{S_i} = y\}, \tag{5}$$

where $Y = \{\text{car, bike, person}\}$, S_i enumerates all segments within an image and y_{S_i} is the label of segment S_i. In words: each pixel is assigned the class with the highest number of class segments containing it. This simple heuristic yields good results in practice.

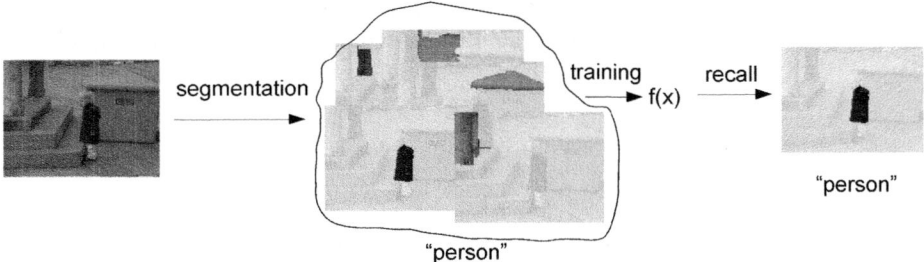

Fig. 1. Overview of our method. See text for details.

4   Experiments

4.1   Instance-Level Predictions Using Multi-instance Kernels

To assess the validity of instance-level predictions using multi-instance kernels, we transform f_I back to an instance-level classifier, using the multi-instance learning assumption (Equation (2)). We refer to these instance-based MIK predictions as MIK-instance. In all experiments, the parameters of the MI-Kernel and SVM are adjusted using MIK and then used with both MIK and MIK-instance. This facilitates very fast parameter scans since MIK is very efficient to compute. Note that we cannot adjust parameters using instance prediction error, as we assume no instance labels to be known.

Table 1. Bag level performance of various MIL algorithms on the standard Musk datasets. All but MIK provide instance-level labeling.

	SVM-SVR	EMDD	mi-SVM	MI-SVM	MICA	MIK	MIK-instance
Musk1	87.9	84.9	87.4	77.9	84.3	88.0	88.0
Musk2	85.4	84.8	83.6	84.3	90.5	89.3	85.2

We compared the performance of MIK, MIK-instance and state-of-the-art MI methods on the Musk benchmark datasets [6], see Table 1. Results were obtained using 10-fold cross-validation. While the computational complexity of MIK-instance is very low compared to the other methods, it achieves competitive results. Using instance-level labels results in a slight loss of accuracy of MIK-instances, compared to MIK. This small degradation of performance is quite surprising, since the model was not trained to provide any instance-level labels.

For multi-class image segmentation, it is beneficial to have a low witness rate, i.e. only a few instances are assumed to be positive in a positive bag. Since an object might not be very prominent in an image, only a fraction of segments might correspond to the object. Table 2 compares the witness rates of MIK-instance, miSVM [1] and SVR-SVM [17] on the Musk datasets. MIK-instance is able to achieve similar accuracy with much less witnesses than the other methods. Note that Musk1 consists of very small bags while Musk2 contains significantly larger bags, more similar to the image/segment setup.

Table 2. MIL algorithms and the empirical witness rates of the classifiers

	Musk1		Musk2	
	accuracy	witness-rate	accuracy	witness-rate
mi-SVM	87.4	100%	83.6	83.9%
SVM-SVR	87.9	100%	85.4	89.5%
MIK-instance	88.0	99%	85.2	62.3%

4.2 Partially Supervised Image Segmentation on Graz 02

We evaluate the performance of the proposed algorithm for object-class segmentation on the challenging Graz-02 dataset. This dataset contains 1096 images of three object classes, bike, car and person. Each image may contain multiple instances of the same class.

We adjusted parameters on a hold-out validation set using bag-level information and used the training and test sets as given by the dataset. It is straightforward to extend the binary MIK method to the multi-class setting using a one-vs-all strategy. We train one MKL-SVM per class using MIK and predict class labels on segment level using MIK-instance. If at least one SVM classifies a segment as positive, it is associated with the most confident class. Otherwise, it is assigned "background" or no class. This yields a classification of each segment into one of four classes: car, bike, person, or background. We merge segments into pixel-level class labels as described in Section 3.4.

Table 3. Pixel-level accuracy on the Graz-02 dataset

	car	bike	person
Segment based MIK-instance (proposed method)	0.30	0.45	0.43
Best strongly supervised approaches [9, 25]	0.72	0.72	0.66

Per-class pixel accuracies are reported in Table 3; some qualitative results are shown in Figure 2. The overall accuracy on images labels, which is the task that was actually trained, is 90%. The performance of our multiple-instance based approach is far from current methods that use pixel-level annotations, whose pixel-level accuracy is around 70% [9, 25] on pixel-level. This is no surprise as our method has no access to the pixel labels. Rather, it is noteworthy that learning segmentation is possible without pixel labels at all.

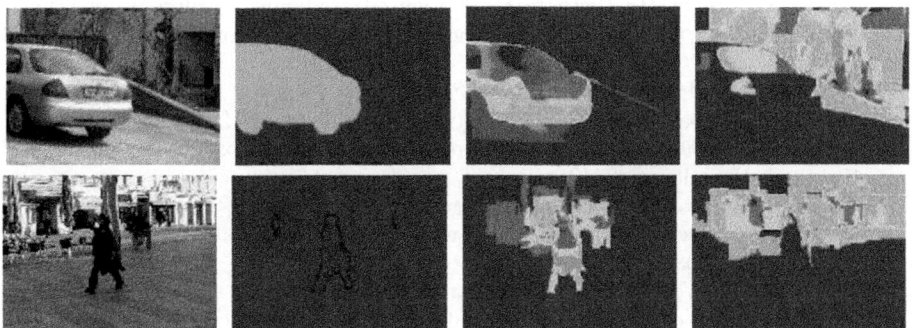

Fig. 2. Qualitative results on on the Graz-02 dataset. Top: Results on category "car". Bottom: Results on category "person". From left to right: original image, ground truth segmentation, segment votes for correct class, segment votes against correct class (red many, blue few votes).

5   Conclusions

We proposed an algorithm for object-class segmentation using only weak supervision based on multiple-instance learning. In our approach, each image is represented as a bag of object-like proposal segments.

We described a way to extend bag-level predictions made by the multi-instance kernel method to instance level while remaining competitive with the state-of-the-art in bag label prediction.

Finally, we evaluated the proposed object-class segmentation method on the challenging Graz02 dataset. While not reaching the performance of methods requiring strong supervision, our result can serve as a baseline for further research into weakly supervised object-class segmentation.

In future work, we plan to scale our approach to much larger image datasets. As much more images with weak annotations are available than with pixel-level segmentation, we hope that we can improve upon the state-of-the-art in object-class segmentation by making use of larger bodies of training images.

References

[1] Andrews, S., Tsochantaridis, I., Hofmann, T.: Support vector machines for multiple-instance learning, pp. 577–584 (2003)
[2] Carreira, J., Sminchisescu, C.: Constrained parametric min-cuts for automatic object segmentation. In: Conference on Computer Vision and Pattern Recognition, pp. 3241–3248 (2010)
[3] Catanzaro, B., Su, B.Y., Sundaram, N., Lee, Y., Murphy, M., Keutzer, K.: Efficient, high-quality image contour detection. In: International Conference on Computer Vision, pp. 2381–2388 (2009)
[4] Chen, Y., Bi, J., Wang, J.Z.: MILES: Multiple-instance learning via embedded instance selection. IEEE Transactions on Pattern Analysis and Machine Intelligence, 1931–1947 (2006)

[5] Dalal, N., Triggs, B.: Histograms of oriented gradients for human detection. In: Converence on Computer Vision and Pattern Recognition, vol. 1, pp. 886–893 (2005)

[6] Dietterich, T., Lathrop, R., Lozano-Pérez, T.: Solving the multiple instance problem with axis-parallel rectangles. Artificial Intelligence 89(1-2), 31–71 (1997)

[7] Endres, I., Hoiem, D.: Category independent object proposals, pp. 575–588. Springer, Heidelberg (2010)

[8] Everingham, M., Van Gool, L., Williams, C.K.I., Winn, J., Zisserman, A.: The Pascal Visual Object Classes (VOC) Challenger. International Journal of Computer Vision 88(2), 303–338 (2010)

[9] Fulkerson, B., Vedaldi, A., Soatto, S.: Class segmentation and object localization with superpixel neighborhoods. In: International Conference on Computer Vision, pp. 670–677 (2009)

[10] Gärtner, T., Flach, P., Kowalczyk, A., Smola, A.: Multi-instance kernels. In: International Conference on Machine Learning, pp. 179–186 (2002)

[11] Gonfaus, J., Boix, X., van de Weijer, J., Bagdanov, A., Serrat, J., Gonzalez, J.: Harmony potentials for joint classification and segmentation. In: Conference on Computer Vision and Pattern Recognition (2010)

[12] Hanbury, A.: How do superpixels affect image segmentation? In: Progress in Pattern Recognition, Image Analysis and Applications, pp. 178–186 (2008)

[13] Jiang, J., Tu, Z.: Efficient scale space auto-context for image segmentation and labeling. In: Conference on Computer Vision and Pattern Recognition, pp. 1810–1817 (2009)

[14] Ladicky, L., Russell, C., Kohli, P., Torr, P.: Associative hierarchical CRFs for object class image segmentation. In: International Conference on Computer Vision, pp. 739–746 (2009)

[15] Leistner, C., Saffari, A., Bischof, H.: MIForests: Multiple-Instance Learning with Randomized Trees. In: Daniilidis, K., Maragos, P., Paragios, N. (eds.) ECCV 2010. LNCS, vol. 6316, pp. 29–42. Springer, Heidelberg (2010)

[16] Li, F., Carreira, J., Sminchisescu, C.: Object recognition as ranking holistic figure-ground hypotheses. In: Conference on Computer Vision and Pattern Recognition, pp. 1712–1719 (2010)

[17] Li, F., Sminchisescu, C.: Convex multiple-instance learning by estimating likelihood ratio. In: Advances in Neural Information Processing Systems. pp. 1360–1368 (2010)

[18] Li, Y.F., Kwok, J., Tsang, I., Zhou, Z.H.: A convex method for locating regions of interest with multi-instance learning. Machine Learning and Knowledge Discovery in Databases. 15–30 (2009)

[19] Lowe, D.: Distinctive image features from scale-invariant keypoints. International Journal of Computer Vision 60(2), 91–110 (2004)

[20] Maire, M., Arbeláez, P., Fowlkes, C., Malik, J.: Using contours to detect and localize junctions in natural images. In: Conference on Computer Vision and Pattern Recognition, pp. 1–8 (2008)

[21] Mangasarian, O., Wild, E.W.: Multiple instance classification via successive linear programming. Journal of Optimization Theory and Applications 137(3), 555–568 (2008)

[22] Nguyen, N.: A New SVM Approach to Multi-instance Multi-label Learning. In: International Conference on Data Mining, pp. 384–392 (2010)

[23] Russell, B., Torralba, A., Murphy, K., Freeman, W.: LabelMe: A database and web-based tool for image annotation. International Journal of Computer Vision 77(1), 157–173 (2008)

[24] Schroff, F., Criminisi, A., Zisserman, A.: Object class segmentation using random forests. In: British Machine Vision Conference (2008)

[25] Schulz, H., Behnke, S.: Object-class segmentation using deep convolutional neural networks. In: Hammer, B., Villmann, T. (eds.) Proceedings of the DAGM Workshop on New Challenges in Neural Computation 2011. Machine Learning Reports, vol. 5, pp. 58–61 (2011)

[26] Van De Sande, K., Gevers, T., Snoek, C.: Evaluating color descriptors for object and scene recognition. Transactions on Pattern Analysis and Machine Intelligence, 1582–1596 (2009)

[27] Verbeek, J., Triggs, B.: Region classification with Markov field aspect models. In: Computer Vision and Pattern Recognition, pp. 1–8 (2007)

[28] Vezhnevets, A., Buhmann, J.: Towards weakly supervised semantic segmentation by means of multiple instance and multitask learning. In: Computer Vision and Pattern Recognition, pp. 3249–3256 (2010)

[29] Vezhnevets, A., Ferrari, V., Buhmann, J.M.: Weakly supervised semantic segmentation with a multi-image model. In: International Conference on Computer Vision (2011)

[30] Wang, S., Siskind, J.: Image segmentation with Ratio Cut. IEEE Transactions on Pattern Analysis and Machine Intelligence, 675–690 (2003)

[31] Zha, Z., Hua, X., Mei, T., Wang, J., Qi, G., Wang, Z.: Joint multi-label multi-instance learning for image classification. In: Conference on Computer Vision and Pattern Recognition, pp. 1–8 (2008)

[32] Zhang, M., Zhou, Z.: M3miml: A maximum margin method for multi-instance multi-label learning. In: International Conference on Data Mining, pp. 688–697 (2008)

[33] Zhang, Q., Goldman, S.A.: Em-dd: An improved multiple-instance learning technique. In: Advances in Neural Information Processing Systems, vol. 2, pp. 1073–1080 (2002)

[34] Zhou, Z., Sun, Y., Li, Y.: Multi-instance learning by treating instances as non-iid samples. In: International Conference on Machine Learning, pp. 1249–1256 (2009)

[35] Zhou, Z., Zhang, M.: Multi-instance multi-label learning with application to scene classification. In: Advances in Neural Information Processing Systems, pp. 1609–1616 (2006)

Semi-supervised Training Set Adaption to Unknown Countries for Traffic Sign Classifiers

Matthias Hillebrand[1], Christian Wöhler[2], Ulrich Kreßel[1], and Franz Kummert[3]

[1] Daimler AG, Group Research and Advanced Engineering, 89081 Ulm, Germany
[2] Image Analysis Group, TU Dortmund, 44221 Dortmund, Germany
[3] Applied Informatics Group, Bielefeld University, 33615 Bielefeld, Germany
{matthias.hillebrand,ulrich.kressel}@daimler.com,
christian.woehler@tu-dortmund.de,
franz@techfak.uni-bielefeld.de

Abstract. Traffic signs in Western European countries share many similarities but also can vary in colour, size, and depicted symbols. Statistical pattern classification methods are used for the automatic recognition of traffic signs in state-of-the-art driver assistance systems. Training a classifier separately for each country requires a huge amount of training data labelled by human annotators. In order to reduce these efforts, a self-learning approach extends the recognition capability of an initial German classifier to other European countries. After the most informative samples have been selected by the confidence band method from a given pool of unlabelled traffic signs, the classifier assigns labels to them. Furthermore, the performance of the self-learning classifier is improved by incorporating synthetically generated samples into the self-learning process. The achieved classification rates are comparable to those of classifiers trained with fully labelled samples.

Keywords: Pattern recognition, self-training, sample selection, confidence bands.

1 Introduction

Many stationary or mobile systems depend on sensory perception of their environment. Intensity-based classifiers are commonly utilised for processing visual sensor information. This contribution considers the automatic recognition of traffic signs by a driver assistance system.

The traffic signs in Western European countries reveal only varieties regarding colour, font, font size and depicted symbols. In a classical supervised learning-based approach, a classifier has to be trained for each country separately. But such an approach is inefficient because of the high labelling costs for human annotators, while unlabelled data can often be acquired done with justifiable expenditure, for example by driving camera-equipped cars and applying an automatic detection algorithm.

A straightforward approach to this problem begins with an initial training set for one country and extends it with the most informative samples from other

F. Schwenker and E. Trentin (Eds.): PSL 2011, LNAI 7081, pp. 120–127, 2012.

countries. An automatic class assignment is desirable to further reduce the labelling costs. This approach is known in literature as semi-supervised learning.

We propose an iterative training process where the most informative samples from a given pool of unlabelled traffic signs are selected by the confidence band method so that a label can be automatically assigned by the classifier itself. Additional knowledge of noise distributions in rotation, camera angle, etc., obtained from the initial training set is provided to the selection algorithm as a set of synthetically generated samples.

The approach evaluation starts with a fully labelled training set of German traffic signs and adapts the system to traffic signs from Austria and Switzerland without any intervention by a human expert. The classifier distinguishes between 12 different classes of traffic signs including speed limits, no-overtaking signs, and the corresponding ending signs. In the following sections, the proposed method is elaborated in detail.

2 Related Work and Applied Methods

2.1 Prerequisites

First, a detection algorithm generates hypotheses from grey-value intensity camera images by applying the Hough method for circles. The second step consists of a normalisation including a resizing of all images to 17×17 pixels and the adjustment of lighting conditions. The intensity values are then used as a feature vector of dimension 289 reduced to 25 by a Principal Component Analysis preserving 81% of the image information.

Since the evaluation only considers traffic sign patterns for the classifier training, the set of hypotheses is divided into a non-sign (garbage) and a sign set in the next step. For this purpose we utilise a second-order polynomial classifier [8] trained on the initial German traffic sign set. The classification leads to a false positive rate of about 5%, which means that each 20th pattern classified as a sign is actually a garbage sample. Similar classification rates for the sign and garbage division on the Austrian and Swiss data sets result from using the classifier trained on the German set.

2.2 Traffic Sign Recognition

Recognition of traffic signs is a mature but still a contemporary field of research. The survey by Fu and Huang [3] provides an introduction and a brief overview of the broad field of existing approaches. Today, traffic sign recognition systems are available as special equipment for some cars of renowned manufacturers. None of these systems have self-learning capabilities.

2.3 Classification

Not all classifiers are equally well-suited for our classification task. The self-learning process nearly always associates a certain fraction of samples with the wrong classes. Different classifier architectures show considerable differences in their sensitivity towards these mislabelled samples. For instance, standard Support Vector Machines (SVM) are very sensitive against outliers and mislabelled training samples [11] while classifiers with a neural network like architecture relying on continuous input, output, and weight values can be trained with partially mislabelled data and are nonetheless capable of producing good classification results [10]. This study utilises a polynomial classifier [8] with a fully quadratic structure of the feature vectors. The decision in favour of this classifier was due to several reasons: First, the training is fast. Second, the classifier is robust to partially mislabelled data. Finally, re-training can be performed easily by mixing old and new moment matrices.

2.4 Self-learning

In the iterative process of training self-learning classifiers, new samples are selected from the large set of available unlabelled samples. The selected samples are then classified or rejected by the classifier, and if not rejected, are added to the training set along with their classifier-predicted labels. At the end of each iteration, the classifier is re-trained with the extended training set and the training procedure is repeated. The surveys by Zhu and Goldberg [12] and Chapelle et al. [1] give a comprehensive overview of the field of semi-supervised learning techniques, also the one described above, commonly referred to as self-training according to Zhu. To the authors' knowledge, self-training methods have not been applied yet to the field of traffic sign recognition.

2.5 Sample Selection

The crucial step in the iterative training process of a semi-supervised or self-learning classifier is the selection of the most informative samples which are to be added to the training set during each iteration in order to re-train the classifier. In self-learning processes without a human teacher, the classifier must be capable of determining labels for the selected unlabelled samples.

An overview to selection methods is given in the survey by Settles [9]. Common approaches are uncertainty sampling [6] and the confidence value estimation from Conditional Random Fields [2]. More recently, the concept of confidence bands was applied to self-learning classifiers for handwritten digits and traffic signs in [4].

2.6 Confidence Bands

Confidence bands are curves enclosing a model function being estimated by a regression analysis. The bands represent the areas where the true model is expected to reside with a certain probability, commonly 95%. The extent of the bands in different areas of the data space gives an idea of how well the estimated function fits the data.

While Martos et al. [7] describe an analytical approach for computing prediction bands in a camera calibration application, Hillebrand et al. [4] adapt the algorithm to compute the closely related confidence bands for application in the context of polynomial classification or regression: A confidence band value is computed for each sample. During each iteration a reference confidence band value (i.e., the average value) is determined from all labelled samples. Based on the minimum difference of their band values to the reference value, a maximum of n samples (e.g., $n = 100$) is selected, classified and added to the training set. The selection of samples with confidence band values close to the average value avoids redundancies (low values) on the one hand and the selection of samples from feature sub-spaces with too much model uncertainty (high values) on the other.

2.7   Virtual Training Samples

Real images of traffic signs depict a wide spectrum of variations, e.g., different sizes, rotations and camera angles, translations due to inaccurate detection results, soiling, partial occlusions, or different lighting and weather conditions. Lighting conditions are normalised by a pre-processing algorithm. The other most common variations (size, rotation, camera angle, translation) are represented by virtual traffic sign samples which are generated from one ideal depiction of each traffic sign by a method described by Hoessler et al. [5], but the less frequent variations are unrepresented. In principle, an infinite amount of such virtual traffic sign samples are available.

3   Experimental Evaluation

3.1   Experimental Setup

The performance evaluation applies the adaptive self-learning classifier to different learning scenarios. As a basis, the classifier is trained with a fully labelled set of German traffic sign samples. This data set consists of 12 classes (see Fig. 1) containing 500 samples per class (6000 samples in total). A training set of the same composition and nearly the same size is available from Austria (5987 samples in total because some classes do not comprise the full number of samples). Furthermore, a smaller training set of only 4428 samples which are not equally distributed over all classes is available from Switzerland (four classes are underrepresented, especially the class *speed limit 70 km/h* with only 11 samples). We refer to the classifier trained with the German training set as *German reference classifier*. The same naming convention applies to the *Austrian* and the *Swiss reference classifier*.

Finally, one set of virtual training samples for each country, again with 12 classes and 6000 samples in total, is created and referred to as the virtual classifiers, e.g., as the *German virtual classifier*.

The classifier performances are compared by computing correct classification rates and false classification rates on independent test sets. Like the training

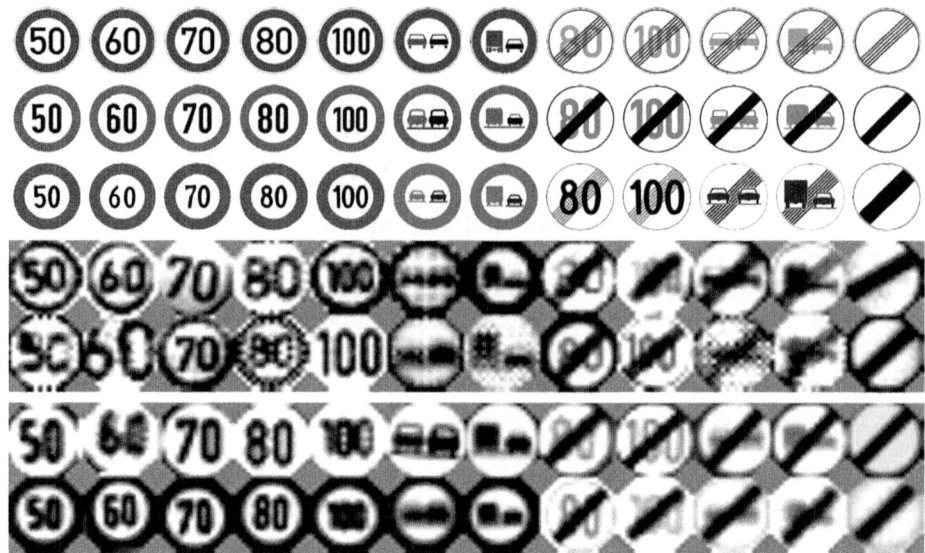

Fig. 1. Traffic sign images. Rows 1–3: Ideal depictions of 12 German, Austrian, and Swiss traffic signs. Rows 4–5: Two German real training samples of each class. Rows 6–7: Two Austrian virtual training samples of each class.

sets, the test sets consist of samples from the same 12 classes. Each class is represented by 250 samples (3000 samples in total) which have been recorded independently with different cameras. The same pre-processing routines have been applied to all samples (real training, virtual training, test).

The selection of German training samples depicted in Fig. 1 shows that the set is not noisy-free and contains a considerable amount of garbage samples and bad quality samples, e.g., with inaccurate cutouts. The proportion of these garbage and bad quality samples is between 5% and 10%. The same applies to the Austrian and Swiss sets. Classifiers often obtain correct classification rates above 90%, so the large number of bad quality samples in the test sets would not allow meaningful comparisons between different classifiers. For that reason, all samples not classifiable by a human expert have been removed from the test sets before the 3000 samples per country were chosen randomly.

For comparison, the performance of the German reference classifier on the German, Austrian, and Swiss test sets as well as the performance of the Austrian reference classifier on the Austrian test set and the Swiss classifier on the Swiss test set were determined. These measurements were repeated with the corresponding virtual classifiers.

As expected, the results presented in Table 1 indicate that the reference classifiers outperform the virtual classifiers in Germany and Austria because some variations described in Section 2.7 are not represented by the virtual training samples. In Switzerland, the virtual classifier performs much better than the

Table 1. Correct classification rates of German, Austrian, and Swiss reference and virtual classifiers on the different test sets. All values are given in percent.

classifier	German		Austrian		Swiss	
	ref.	virt.	ref.	virt.	ref.	virt.
German test set	97.4	89.9				
Austrian test set	88.1	77.8	96.3	87.7		
Swiss test set	85.3	73.1			89.2	95.4

reference classifier due to missing training samples for some classes and the high number of bad quality samples.

3.2   Self-learning with Real Traffic Signs

The German reference classifier performs 88.1% on the Austrian and 85.3% on the Swiss test set; the performances of the Austrian and Swiss virtual classifiers amount to 87.7% and 95.4%, respectively. The objective is for the generated self-learning classifiers to have a higher recognition performance.

The training process starts with an initial set containing all fully labelled German training samples. Then the training continues iteratively as described by adding Austrian samples which have been labelled by the classifier itself. Our classifier outperforms the German reference classifier (88.1%) after some iterations and once all training samples have been added, a performance of 93.1% is reached.

The difference to the Austrian reference classifier (using the same training samples) can be explained by the presence of garbage samples in the training set (about 5% as described in Section 2.1) that cannot be assigned a "correct" class label. Furthermore, a certain fraction of the added samples (about 15%) are mislabelled by the classifier and influence the training process negatively, especially when appearing in the early stages of the process. Making use of both of these samples in self-training will inevitably twist the feature distribution of each class.

A completely different behaviour can be observed when training the Swiss self-learning classifier: the performance (81.5%) is even lower than the performance of the German reference classifier (85.3%). This is likely due to the missing training samples and the bad quality of the existing ones, which results in a high fraction of mislabelled samples (around 31% of the added samples).

Further experiments vary the initial training set sizes by adding 10% (600 samples) of the labelled Austrian and Swiss samples, respectively. As a result, the self-learning classifiers obtain much better performances due to a lower rate of mislabelled samples (around 15% and 29%, respectively): the Austrian self-learning classifier improves to 95.3% and the Swiss classifier to 84.4%. Of course, the performance increases come at the price of being dependent on a human labelling expert to some extent again.

Adding another supplemental 10% labelled Austrian samples (now 1200 samples in total) to the initial training set results in a decrease of the rate of mislabelled samples by another 0.5% but has nearly no improving effect on the classification rates. With the additional 10% labelled samples, the Swiss classifier improves marginally to 85.2% and the rate of mislabelled samples decreases to 27%, but this performance is still worse than that of the German reference classifier.

3.3   Self-learning with Virtual Traffic Signs

When disposing of a theoretically endless supply of virtual training samples, it appears reasonable to extend the German standard classifier with a huge amount of Austrian respectively Swiss virtual training samples. After adding 6000 virtual training samples each, classifier performances of 95.2% (Austrian) and 94.9% (Swiss) are reached. Clearly, the classification rate of the Austrian classifier is nearly equal to the one of the self-learning classifier trained before, while the performance of the Swiss classifier is nearly equal to that of the Swiss virtual classifier.

Finally, the self-learning classifiers are combined with the recently trained virtual classifiers. The initial training sets are constructed from all 6000 German fully labelled samples and all 6000 Austrian and Swiss virtual samples, respectively, also fully labelled. Then we start the self-learning process by adding all Austrian respectively Swiss real training samples iteratively. These classifiers perform with correct classification rates of 95.7% and 95.2%, respectively, which are the best performances except for the Austrian reference classifier and the Swiss virtual classifier. The remarkable point here is that these high performances were reached without any intervention by a human labelling expert.

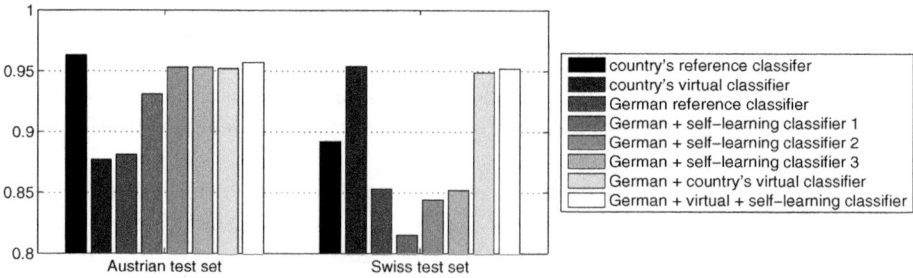

Fig. 2. Classification rates. Comparison of the correct classification rates of the classifiers described in Sections 3.2 and 3.3 on the Austrian and the Swiss training set.

4   Summary and Conclusions

The proposed self-learning classification system is capable of adapting itself to changed appearances of known traffic signs from other countries. This reduces the labelling efforts by human expert annotators and therefore the overall costs significantly.

The Swiss classifier achieves its best performance when trained with Swiss virtual samples only. This behaviour is due to the fact that the Swiss training set contains a huge amount of bad quality samples. A comparison with other countries reveals that a classifier exclusively trained based on virtual samples is always capable of classifying traffic sign images to a certain degree correctly.

The Austrian classifier achieves the best performance when trained with virtual samples first and subsequently self-trained with real samples. Since a Swiss classifier trained in this way performs nearly as good as its virtual classifier, this strategy is a suitable compromise.

References

1. Chapelle, O., Schölkopf, B., Zien, A. (eds.): Semi-Supervised Learning. Adaptive Computation and Machine Learning. The MIT Press (2006)
2. Culotta, A., McCallum, A.: Confidence Estimation for Information Extraction. In: Proc. of Human Language Technology Conference and North American Chapter of the Association for Computational Linguistics (HLT-NAACL), pp. 109–112 (2004)
3. Fu, M.Y., Huang, Y.S.: A survey of traffic sign recognition. In: Proc. of the International Conference on Wavelet Analysis and Pattern Recognition (ICWAPR), pp. 119–124 (2010)
4. Hillebrand, M., Wöhler, C., Krüger, L., Kreßel, U., Kummert, F.: Self-learning with confidence bands. In: Proc. of the 20th Workshop Computational Intelligence, pp. 302–313 (2010)
5. Hoessler, H., Wöhler, C., Lindner, F., Kreßel, U.: Classifier training based on synthetically generated samples. In: Proc. of the 5th International Conference on Computer Vision Systems (ICCV) (2007)
6. Jeon, J.H., Liu, Y.: Semi-supervised Learning for Automatic Prosodic Event Detection Using Co-training Algorithm. In: Proc. of the 47th Annual Meeting of the ACL and the 4th IJCNLP of the AFNLP, pp. 540–548 (2009)
7. Martos, A., Krüger, L., Wöhler, C.: Towards Real Time Camera Self Calibration: Significance and Active Selection. In: Proc. of the 4th Int. Symp. on 3D Data Processing, Visualization and Transmission (3DPVT) (2010)
8. Schürmann, J.: Pattern Classification: A Unified View of Statistical and Neural Approaches. John Wiley & Sons (1996)
9. Settles, B.: Active Learning Literature Survey. Computer Sciences Technical Report 1648. University of Wisconsin–Madison (2010)
10. Wöhler, C.: Autonomous in situ training of classification modules in real-time vision systems and its application to pedestrian recognition. Pattern Recognition Letters 23(11), 1263–1270 (2002)
11. Xu, L., Crammer, K., Schuurmans, D.: Robust Support Vector Machine Training via Convex Outlier Ablation. In: Proc. of the 21st National Conference on Artificial Intelligence (AAAI), pp. 536–542 (2006)
12. Zhu, X., Goldberg, A.B.: Introduction to Semi-Supervised Learning. Synthesis Lectures on Artificial Intelligence and Machine Learning. Morgan & Claypool Publishers (2009)

Comparison of Combined Probabilistic Connectionist Models in a Forensic Application

Edmondo Trentin[1], Luca Lusnig[1], and Fabio Cavalli[2]

[1] Dip. di Ingegneria dell'Informazione, Università di Siena, Italy
[2] Research Unit of Paleoradiology and All. Sci., AOUTS Trieste, Italy
trentin@dii.unisi.it

Abstract. A growing interest toward automatic, computer-based tools has been spreading among forensic scientists and anthropologists wishing to extend the armamentarium of traditional statistical analysis and classification techniques. The combination of multiple paradigms is often required in order to fit the difficult, real-world scenarios involved in the area. The paper presents a comparison of combination techniques that exploit neural networks having a probabilistic interpretation within a Bayesian framework, either as models of class-posterior probabilities or as class-conditional density functions. Experiments are reported on a severe sex determination task relying on 1400 scout-view CT-scan images of human crania. It is shown that connectionist probability estimates yield higher accuracies than traditional statistical algorithms. Furthermore, the performance benefits from proper mixtures of neural models, and it turns up affected by the specific combination technique adopted.

Keywords: Multiple classifier, neural net, density estimation, forensics.

1 Introduction

In recent times, a growing interest toward automatic, computer-based tools has been spreading among forensic scientists and anthropologists wishing to extend the armamentarium of traditional statistical analysis and classification techniques [8]. In particular, reliable methods for the determination of the sex from human skeletal remains is of fundamental importance, for identification in forensic cases and for paleodemographic studies on ancient populations [2]. The sexual dimorphism is better recognizable in the pelvis, but (because of its complex shape) the latter is often found in very poor condition. A fundamental alternative is thus represented by the skull, which is generally better preserved and more easily reconstructed if found fragmented [7]. The paper copes with sex classification from scout-view computerized tomography (CT)-scan images of male and female human skulls, relying on 1400 images collected on the field. In particular, the goal is twofold: (i) searching for a reliable solution to the problem, applying either statistical or neural network approaches within a Bayesian framework; (ii) investigating and comparing different techniques for combining connectionist estimates of the probabilistic quantities involved in the maximum-a-posteriori classification strategy.

F. Schwenker and E. Trentin (Eds.): PSL 2011, LNAI 7081, pp. 128–137, 2012.

As reviewed in Section 2, a probabilistic interpretation of the output of a neural network can be given in terms of a supervised, discriminative posterior-probability setup, or in terms of unsupervised class-conditional density estimation. While the former is the traditional practitioner's choice in pattern recognition applications of neural networks, the latter is far less investigated in the literature, mostly due to the intrinsic difficulties which arise in dealing with the unsupervised estimation task. Nonetheless, robust class-conditional density estimates can be used *per se* within Bayes theorem as viable classification tools. Moreover, they can capture and convey relevant information that can be combined with the class-posterior estimates in order to improve the performance of the overall multiple-classifier system. To this end, we rely on a neural network approach to the density estimation task that we proposed in [10] (reviewed in Section 2, as well). Note that in [10] the experimental evaluation of the model was carried out on illustrative, univariate synthetic datasets generated with probability density functions (pdf) having known form. Therefore, an additional aim of this paper is the evaluation of the approach in a multivariate, real-world task.

The combination techniques evaluated in the paper are presented in Section 3. They rely on two common, somewhat complementary notions. First, having models of probabilistic quantities may ease the definition of meaningful combination schemes that benefit from the homogeneous nature of the underlying classifiers (possibly, turning themselves out to undergo a plausible interpretation in terms of probabilities). Second, on the other way around, posterior probability models and individual class-conditional density functions are the carrier of non-completely overlapping information, providing the combination algorithm with the opportunity to perform better than the separate models actually do. Sex determination experiments on an original, real-scale dataset are reported in Section 4. Some conclusions, relevant to the machine learning as well as to the anthropology/forensic sciences communities are drawn in Section 5.

2   Probabilistic Interpretation of Neural Networks

Artificial neural networks (ANNs) [4,1] have been widely applied to pattern classification tasks [1]. In most cases, their application takes the form of a connectionist discriminant function, which is trained to yield a high "score" on the correct class, along with low scores on all the wrong classes. No probabilistic interpretation of such a discriminant function is usually given, neither it is even expected. As a matter of fact, minimum classification error is gained when a maximum class-posterior probability is chosen as a discriminant within a Bayesian framework [3]. This is accomplished relying on the popular Bayes theorem[3], i.e. $P(\omega_i \mid \mathbf{x}) = p(\mathbf{x} \mid \omega_i)P(\omega_i)/p(\mathbf{x})$, where $\mathbf{x}$ is a pattern (real-valued feature vector) to be assigned to one out of c distinct and disjoint classes $\omega_1, \ldots, \omega_c$. The theorem transforms a prior knowledge on the probability of individual classes, i.e. the prior probability $P(\omega_i)$, into a posterior knowledge upon observation of a certain feature vector $\mathbf{x}$, namely the posterior probability

$P(\omega_i \mid \mathbf{x})$. Such a transformation relies on the evaluation of the so-called class-conditional pdf $p(\mathbf{x} \mid \omega_i)$. Theorems confirm that, under rather mild conditions, ANNs can be trained as optimal estimates of Bayes posterior probabilities [1]. These theorems give a mathematical foundation to the popular heuristic decision rules that we mentioned at the beginning of this section. Roughly speaking, it can be shown that a multi-layer perceptron (MLP) [4] having c output units and trained via regular backpropagation (BP) [4] over a labeled training set $\mathcal{T} = \{(\mathbf{x}_k, \mathbf{y}_k) \mid k = 1, \ldots, n\}$ where $\mathbf{y}_k = (y_{k1}, \ldots, y_{kc})$ and $y_{ki} = \begin{cases} 1 \text{ if } \mathbf{x}_k \in \omega_i \\ 0 \text{ otherwise} \end{cases}$ is an "optimal" non-parametric estimation of the left-hand-side of Bayes theorem. In practice, it is not necessary to know the class-posterior probabilities in advance in oder to create target outputs for the BP training, since a crisp 0/1 labeling (which reminds us of the good, old Widrow-Hoff labeling for linear discriminant [3]) drives the ANN weights to convergence towards the same result. Since a probabilistic interpretation of the MLP outputs is sought, some constraints are required. First, output values are limited to the $(0, 1)$ range. This is readily accomplished by relying on the usual sigmoid activation functions. Then, since $\sum_{i=1}^{c} P(\omega_i \mid \mathbf{x}) = 1$, a normalization of the MLP outputs is needed.

Whilst estimation of posterior probabilities via ANNs is feasible due to the simplicity of satisfying the probability constraints, connectionist estimation of pdfs–i.e., class-conditional pdfs to be used in the right-hand-side of Bayes Theorem–is much harder, since: (i) a pdf may possibly take any non-negative, unbounded value; (ii) its integral over the feature space shall equal 1; (iii) above all, pdf estimation is an intrinsically unsupervised learning problem, and standard training algorithms do not do. Yet, due to their flexibility and generalization capabilities, neural models of pdfs could improve over parametric and non-parametric statistical estimation techniques. In [10] we proposed a connectionist model for density estimation which overcomes the major limitation of statistical techniques. A concise review of the approach follows. Let us consider a pdf $p(\mathbf{x})$, defined over a real-valued, d-dimensional feature space. The model is introduced along the usual line followed in the traditional kernel-based nonparametric pdf estimates, such as the Parzen window (PW) [3]. These techniques are built on the observation that the probability that a pattern $\mathbf{x}' \in \mathcal{R}^d$, drawn from $p(\mathbf{x})$, falls in a certain region R of the feature space is $P = \int_R p(\mathbf{x})d\mathbf{x}$. Let then $\mathcal{T} = \{\mathbf{x}_1, \ldots, \mathbf{x}_n\}$ be an unsupervised sample of n patterns, identically and independently distributed (i.i.d.) according to $p(\mathbf{x})$. If k_n patterns in $\mathcal{T}$ fall within R, an empirical estimate of P can be obtained as $P \simeq k_n/n$. If $p(\mathbf{x})$ is continuous and R is small enough to prevent $p(\mathbf{x})$ from varying its value over R in a significant manner, we are also allowed to write $\int_R p(\mathbf{x})d\mathbf{x} \simeq p(\mathbf{x}')V$, where $\mathbf{x}' \in R$, and V is the volume of region R. An estimated value of the pdf $p(\mathbf{x})$ over pattern $\mathbf{x}'$ is thus given by:

$$p(\mathbf{x}') \simeq \frac{k_n/n}{V_n} \tag{1}$$

where V_n denotes the volume of region R_n, assuming that smaller regions around $\mathbf{x}'$ are considered as the sample size n increases. This is expected to allow equation (1) to yield improved estimates of $p(\mathbf{x})$, i.e. to converge to the exact value of $p(\mathbf{x}')$ as n (hence, also k_n) tends to infinity (a discussion of the asymptotic behavior of nonparametric models of this kind can be found in [3]). The basic instance of the PW technique assumes that R_n is a hypercube having edge h_n, such that $V_n = h_n^d$. The edge h_n is usually defined as a function of n as $h_n = h_1/\sqrt{n}$, in order to ensure a correct asymptotic behavior. The value h_1 has to be chosen empirically, and it heavily affects the resulting model. The formalization of the idea requires to define a unit-hypercube window function in the form $\varphi(\mathbf{y}) = \begin{cases} 1 & \text{if } | y_j | \leq 1/2, j = 1, \ldots, d \\ 0 & \text{otherwise} \end{cases}$, such that $\varphi(\frac{\mathbf{x}'-\mathbf{x}}{h_n})$ has value 1 iff $\mathbf{x}'$ falls within the d-dimensional hyper-cubic region R_n centered in $\mathbf{x}$ and having edge h_n. This implies that $k_n = \sum_{i=1}^n \varphi(\frac{\mathbf{x}'-\mathbf{x}_i}{h_n})$. Using this expression, from equation (1) we can write

$$p(\mathbf{x}') \simeq \frac{1}{n} \sum_{i=1}^n \frac{1}{V_n} \varphi(\frac{\mathbf{x}' - \mathbf{x}_i}{h_n}) \tag{2}$$

which is the PW estimate of $p(\mathbf{x}')$ from the sample $\mathcal{T}$. The model is usually refined by considering smoother window functions $\varphi(.)$, instead of hypercubes. The idea for training a MLP to estimate $p(\mathbf{x})$ from $\mathcal{T}$ is to use the PW model as a target output for the ANN, and to apply standard BP to the MLP. A unbiased variant of this idea is proposed, according to the following unsupervised algorithm (expressed in pseudo-code):

```
Input: T = {x₁,...,xₙ}, h₁.
Output: p̃(.) /* the connectionist estimate of p(.) */
```

1. Let $h_n = h_1/\sqrt{n}$
2. Let $V_n = h_n^d$
3. For i=1 to n do /* loop over $\mathcal{T}$ */
3.1   Let $\mathcal{T}_i = \mathcal{T} \setminus \{\mathbf{x}_i\}$
3.2   Let $y_i = \frac{1}{n-1} \sum_{\mathbf{x} \in \mathcal{T}_i} \frac{1}{V_{n-1}} \varphi(\frac{\mathbf{x}_i - \mathbf{x}}{h_{n-1}})$ /* target output */
4. Let $\mathcal{S} = \{(\mathbf{x}_i, y_i) \mid i = 1, \ldots, n\}$ /* supervised training set */
5. Train the ANN via BP over $\mathcal{S}$
6. Let $\tilde{p}(.)$ be the function computed by the ANN
7. Return $\tilde{p}(.)$

Since the ANN output is assumed to be an estimate of a pdf, it must be nonnegative, yet unbounded. For this reason, sigmoids with adaptive amplitude λ (i.e., in the form $y = \frac{\lambda}{1+e^{-x}}$), as described in [9], are used as output activation functions. As in several statistical nonparametric models, such as the k_n-nearest neighbor technique [3], the ANN is not necessarily a pdf (in general, the integral of $\tilde{p}(.)$ over the feature space is not 1), but a good (i.e., useful) approximation of the desired density is obtained, overcoming the limitations of traditional estimation methods [10]. We refer to this model as the Parzen-ANN (P-ANN). In this paper, the P-ANN is applied to the estimation of the class-conditional density functions $p(\mathbf{x} \mid \omega_i)$ to be used within Bayes theorem. This means that individual, class-specific networks have to be trained over the data belonging to the corresponding class. Standard Gaussian kernels will be applied in the experiments (step 3.2 of the algorithm).

3   Combination Techniques

The probabilistic interpretation of different neural models provides us with a number of simple yet well-grounded combination techniques for a multiple classifier system. For notational convenience, for each class $i = 1, \ldots, c$ we write $\hat{P}(\omega_i \mid \mathbf{x})$ to denote the posterior estimate of $P(\omega_i \mid \mathbf{x})$ yielded by the i-th output of the supervised MLP, and $\tilde{P}(\omega_i \mid \mathbf{x})$ to refer to the quantity $\tilde{p}(\mathbf{x} \mid \omega_i)P(\omega_i)/\tilde{p}(\mathbf{x})$, where $\tilde{p}(\mathbf{x} \mid \omega_i)$ is the P-ANN for the class-conditional $p(\mathbf{x} \mid \omega_i)$ and $\tilde{p}(\mathbf{x})$, the estimate of the evidence $p(\mathbf{x})$, is obtained as $\sum_{j=1}^{c} P(\omega_j)\tilde{p}(\mathbf{x} \mid \omega_j)$, as usual. Plausible combination techniques may be defined as follows.

1. *Pseudo-joint probability:* let ξ_1 and ξ_2 be the random quantities yielded by two distinct functions (or, regression models) of a given random vector $\mathbf{x} \in \Re^d$. We refer to ξ_1 and ξ_2 as the "models", and the following discussion can be extended straightforwardly to an arbitrary number of models. For any generic state of nature ω_i, $i = 1, \ldots, c$, we can write:

$$P(\omega_i \mid \xi_1, \xi_2) = \frac{p(\xi_1, \xi_2 \mid \omega_i)P(\omega_i)}{p(\xi_1, \xi_2)} \tag{3}$$
$$= \frac{p(\xi_1 \mid \omega_i)p(\xi_2 \mid \xi_1, \omega_i)P(\omega_i)}{p(\xi_1, \xi_2)}.$$

Under the assumption that the models are independent of each other, equation (3) can be rewritten as follows:

$$P(\omega_i \mid \xi_1, \xi_2) = \frac{p(\xi_1 \mid \omega_i)p(\xi_2 \mid \omega_i)P(\omega_i)}{p(\xi_1)p(\xi_2)} \tag{4}$$
$$= \frac{P(\omega_i \mid \xi_1)p(\xi_1)}{p(\xi_1)P(\omega_i)} \frac{P(\omega_i \mid \xi_2)p(\xi_2)}{p(\xi_2)P(\omega_i)} P(\omega_i)$$
$$= \frac{P(\omega_i \mid \xi_1)P(\omega_i \mid \xi_2)}{P(\omega_i)}$$

which has the form of a pseudo-joint probability (the product of quantities at the numerator) normalized by the class-prior. The use of the expression "pseudo" is enforced by the observation that in real-world scenarios the models are hardly independent, yet equation (4) can still be fruitfully applied in a naive-Bayes fashion. If the classes are equally alike a priori (as in the experiments reported in the paper), i.e. if $P(\omega_i) = P(\omega_j)$ for each $i, j \in$

$\{1, \ldots, c\}$, then a discriminant function $g_i(.)$ can be defined for each class ω_i by taking the usual maximum-a-posteriori probability given the models, i.e. $\max_i P(\omega_i \mid \xi_1, \xi_2)$, and dropping the denominator from Eq. (4). In so doing, discriminant functions are defined as pseudo-joint probabilities in the form $g_i(\mathbf{x}) = P(\omega_i \mid \xi_1(\mathbf{x}))P(\omega_i \mid \xi_2(\mathbf{x}))$, and the corresponding decision rule assigns a pattern $\mathbf{x}$ to class i if $g_i(\mathbf{x}) \geq g_j(\mathbf{x})$ for each $j \neq i$, as usual. In the experiments we assume that $\xi_1(.)$ is the supervised MLP and $\xi_2(.)$ is realized via P-ANN (and Bayes theorem), and we let $P(\omega_i \mid \xi_1(\mathbf{x})) \approx \hat{P}(\omega_i \mid \mathbf{x})$ and $P(\omega_i \mid \xi_2(\mathbf{x})) \approx \tilde{P}(\omega_i \mid \mathbf{x})$, according to the notation above.

2. *Maximum confidence*: when we assign a pattern $\mathbf{x}$ to class ω_i according to the maximum-a-posteriori criterion, i.e. $i = argmax_j P(\omega_j \mid \mathbf{x})$, we face a certain Bayesian risk, namely the probability of misclassification given the pattern. The latter can be written as $P(error \mid \mathbf{x}) = \sum_{j=1, j \neq i}^{c} P(\omega_j \mid \mathbf{x})$. It is seen that the higher the posterior probability of ω_i, the lower the probability of error. In the present setup, a minimum-risk combination strategy for the two connectionist models follows in a natural manner: if the neural networks agree on the decision of assigning pattern $\mathbf{x}$ to ω_i, just do it. Otherwise, if $\hat{P}(\omega_i \mid \mathbf{x}) \geq \hat{P}(\omega_j \mid \mathbf{x})$ for all $j \neq i$ and $\tilde{P}(\omega_k \mid \mathbf{x}) \geq \tilde{P}(\omega_j \mid \mathbf{x})$ for all $j \neq k$, then the decision $d(\mathbf{x})$ between ω_i and ω_k is taken as:

$$d(\mathbf{x}) = \begin{cases} \omega_i & \text{if } \hat{P}(error \mid \mathbf{x}) \geq \tilde{P}(error \mid \mathbf{x}) \\ \omega_k & \text{otherwise} \end{cases} \tag{5}$$

where $\hat{P}(error \mid \mathbf{x}) = \sum_{j=1, j \neq i}^{c} \hat{P}(\omega_j \mid \mathbf{x})$ and $\tilde{P}(error \mid \mathbf{x}) = \sum_{j=1, j \neq k}^{c} \tilde{P}(\omega_j \mid \mathbf{x})$. In other words, the classification relies eventually on the model which exhibits the highest confidence in its own decision.

3. *Minimum expectation*: albeit appealing, the combination based on maximum confidence has a major drawback. In fact, a rough model of the Bayesian posterior probability turns implicitly out to be a rough estimator of its own Bayesian risk, as well (e.g., by over-estimating the class-posterior over a certain pattern, resulting in an under-estimate of the corresponding probability of error). This may suggest taking a somewhat complementary approach, discarding the (overwhelmingly optimistic) maximum-confidence decision and opting for the (possibly more realistic) minimum expectation strategy. In this framework, the latter takes the following form: if the two models are in disagreement, say $d(\mathbf{x}) = \omega_i$ based on $\hat{P}(\omega_i \mid \mathbf{x})$ and $d(\mathbf{x}) = \omega_k$ based on $\tilde{P}(\omega_k \mid \mathbf{x})$, then assign $\mathbf{x}$ to ω_i if $\hat{P}(\omega_i \mid \mathbf{x}) \leq \tilde{P}(\omega_k \mid \mathbf{x})$, else assign $\mathbf{x}$ to ω_k. Albeit heuristic, this conservative strategy reveals to be backed up by empirical evidence.

4. *Average*: a natural, simple alternative is represented by the average between the two estimates, namely taking $P(\omega_i \mid \mathbf{x}) \approx \frac{1}{2}\hat{P}(\omega_i \mid \mathbf{x}) + \frac{1}{2}\tilde{P}(\omega_i \mid \mathbf{x})$ for each $i = 1, \ldots, c$. The straightforward extension of the technique relies on a *weighted average* over the models in the form $P(\omega_i \mid \mathbf{x}) \approx \alpha\hat{P}(\omega_i \mid \mathbf{x}) + (1 - \alpha)\tilde{P}(\omega_i \mid \mathbf{x})$ for each class, where the relative weight $\alpha \in (0, 1)$ can be determined empirically via model selection techniques, contributing to compensate for possible biases and/or numeric mismatches between the two models.

5. *Rejection on ξ_ι*: a variation on the theme of the maximum confidence, which outsprings from the same background reasoning and from a long-standing tradition in practical development of classifiers which include the reject option (i.e. reject current pattern $\mathbf{x}$, refusing to take any decision, whenever the estimated value of the discriminant functions $g_1(\mathbf{x}), \ldots, g_c(\mathbf{x})$ are all below a given rejection threshold θ, with θ in the $(0,1)$ interval) can be introduced as follows. Let $\xi_1(\mathbf{x}) = \hat{P}(\omega_i \mid \mathbf{x})$ and $\xi_2(\mathbf{x}) = \tilde{P}(\omega_k \mid \mathbf{x})$, where ω_i and ω_k are the decisions taken by models ξ_1 and ξ_2 over $\mathbf{x}$, respectively. We say that a rejection on ξ_1 decision strategy assigns $\mathbf{x}$ to ω_i if $\xi_1(\mathbf{x}) \geq \theta$, and to ω_k otherwise (regardless of the value of $\xi_2(\mathbf{x})$). On the other way around, the rejection on ξ_2 assigns by default to ω_k, unless $\xi_2(\mathbf{x}) < \theta$ (in the latter case $\mathbf{x}$ is assigned to ω_i). It is seen that these decision rules do not coincide with the maximum-confidence approach. Suitable values for θ are found empirically, within a proper model selection framework.

6. *Mixture of experts*: in principle, the most flexible combination technique simply avoids arbitrary choices on the explicit combination strategy, and lets the machine learn its own "optimal" recipe from examples. A straightforward, yet sound realization of this principle relies on a committee of neural experts [4]. In the present setup we consider a third MLP which, for each pattern $\mathbf{x}$, is fed with the estimates $\xi_1(\mathbf{x})$ and $\xi_2(\mathbf{x})$ and is expected to yield in output a more robust estimate of $P(\omega_i \mid \mathbf{x})$. We refer to this third connectionist module as the gating network. More precisely, in a c-class problem $\xi_1(\mathbf{x})$ has c output units, forming an input vector $(\hat{P}(\omega_1 \mid \mathbf{x}), \ldots, \hat{P}(\omega_c \mid \mathbf{x}))$ while $\xi_2(\mathbf{x})$ is better described as an ordered collection of c separate P-ANNs, say $(\tilde{p}(\mathbf{x} \mid \omega_1), \ldots, \tilde{p}(\mathbf{x} \mid \omega_c))$. The aggregate vector $(\hat{P}(\omega_1 \mid \mathbf{x}), \ldots, \hat{P}(\omega_c \mid \mathbf{x}), \tilde{p}(\mathbf{x} \mid \omega_1), \ldots, \tilde{p}(\mathbf{x} \mid \omega_c))$ defines the input space for the gating network, whose target output is the usual, Widrow-Hoff-like binary coding $(0/1)$ of the correct class whom the current training pattern belongs to. In so doing, as remarked in Section 2, the gating network approximates the Bayesian class-posterior probability, learning the combination law of its inputs which best fits its training criterion. To practical ends, $\xi_1(\mathbf{x})$ and $\xi_2(\mathbf{x})$ are separately trained first, as usual. Later on, the gating network is trained (with regular BP) on the outputs yielded by $\xi_1(\mathbf{x})$ and $\xi_2(\mathbf{x})$ over the original training data.

4   Experiments

For this study, a total of 1400 scout-view CT scanogram (of healthy, adult, Caucasian subjects) were selected at random from our PACS database, including 700 males and 700 females within an age range of 25–92. The scanogram was chosen because it is routinely performed before a cranial CT examination, and since for our purposes (i.e., the determination of the external shape of calvarium in norma lateralis) it is basically as reliable as the cephalometric lateral radiograph. The patients were chosen on the basis of their residence in the province of Trieste (Italy), since the population of this geographic area is the result of complex historical genetic crossover between Italic, Germanic and Slavic populations. Lateral cranial scanograms were automatically selected and anonymized by our PACS facilities (registering only the sex and the age) among the CT examinations performed between the years 2005 and 2010 in the radiological structures of the Department of Diagnostic Imaging of the Hospital Corporation at the University of Trieste with similar multislice computed tomography (MSCT) equipment. Lateral CT scanograms were taken on an Aquilion 16 Toshiba multislice CT scanner, using the standard preset (120 kVp, 150 mAs, matrix size 512x512). The images were automatically transformed from DICOM to JPG format, maintaining the original matrix size.

Visual feature extraction from the images underwent the following procedure. A smoothing Gaussian filter (with discrete Weierstrass transform relying on a 5×5 convolution matrix) is applied first [6], in order to reduce additive noise. It is followed by a sharpening filter. Starting from the filtered image, edge detection and edge connection are accomplished by a technique relying on Canny algorithm, followed by thresholding. Upon removal of the maxilla and mandible area, the contour of the *cranium* is extracted automatically (including the *glabella, calvarium*, and *opisthion* areas). The centroid-distance signature function is then extracted [12], ensuring translation-invariance. In order to reduce the dimensionality significantly, and to resort to a fixed-dimensionality representation, sub-sampling of the overall set of signatures is accomplished via the equal points sampling technique. Features are finally extracted from the sub-sampled signatures by application of the usual fast Fourier transform (FFT), retaining the first 64 parameters. This results in a 64-dim feature space which ensures rotation invariance and scale invariance (by proper normalization of the magnitude of the first half of the FFT coefficients), as described in [12].

The data were split first into a training and a validation set, for model selection purposes. Once the selection process was completed and upon replacement of the original data, the patterns were then randomly partitioned again into a training set (1000 patterns), and a test set (400 patterns), having an equal balance between the relative frequencies of male and female samples. Results are reported in Table 1 (the same notation used in the previous section is used to refer to the specific models). Linear discriminant analysis was applied first (applying the pseudo-inverse method based on singular value decomposition), in order to fix a baseline. The results confirm the high-nonlinearity of the classification task. A more significant baseline was yielded by a regular k-nearest neighbor (k-NN) classifier with $k = 5$. The performance turnt out to be improved by the PW approach. Standard Gaussian kernels were used, with initial width $h_1 = 9.77 \times 10^{-2}$. Connectionist approaches follow, starting from the individual classifiers relying on unsupervised estimation of $p(\mathbf{x} \mid \omega_i)$ (15 hidden sigmoid units for the "male" class, and 16 such units for the "female" class; sigmoid activation in the output unit, all activation functions having a smoothness set to 0.4 and layer-by-layer adaptive amplitude). Training these P-ANNs required 300 epochs only, with learning rates $\eta = 0.1$. The next row of the table shows the results yielded by the supervised estimation of $P(\omega_i \mid \mathbf{x})$ via MLP. The latter has 16 hidden sigmoid units and a sigmoid

Table 1. Sex recognition rate using the cranium contour

Model	Accuracy (%)
Linear discriminant	53.80
k-NN	68.25
Parzen Window	70.75
$\tilde{P}(\omega_i \mid \mathbf{x})$	79.25
$\hat{P}(\omega_i \mid \mathbf{x})$	80.25
Pseudo-joint probability	82.00
Maximum confidence	81.50
Minimum expectation	81.25
Average	82.00
Weighted average	82.75
Rejection on ξ_1	83.00
Rejection on ξ_2	81.75
Mixture of experts	83.50

output, all having smoothness 1.25. 20000 epochs of BP with learning rate $\eta = 0.1$ were applied. Accuracies turn out to outperform the statistical techniques. The P-ANN performance is even surprisingly higher than the traditional PW, and close to the supervised, discriminative MLP. The combination techniques proposed in Section 3 are reported in the next rows of the table. The mixture of experts relies on a gating MLP with 9-hidden sigmoid units and a sigmoid output (all smoothnesses set to 1), and neuron-by-neuron adaptive amplitudes. Training required 150 epochs with a learning rate set to 0.02. It is seen that all the combination methods are effective (although, with a certain variance in terms of relative performance), showing that the difference in the information conveyed by the connectionist models involved are complementary to some extent and can be exploited jointly in order to come up with a more robust classifier. Letting the machine discover the most suitable combination law (relying on the committee machine) yields higher recognition rates than fixed (albeit plausible) mixing choices. In the best case scenario (i.e., mixture of experts) a relative error rate reduction of 16.46% is gained w.r.t the best single-model classifier. Results are of the utmost significance in an application oriented perspective, if compared with the expected recognition rate ($\sim 80\%$) by human experts [11], as well as similar classification experiments carried out using statistical approaches in the forensic sciences [5].

5   Conclusions

The paper faced a difficult, real-world classification task having the utmost relevance to anthropology and the forensic sciences, namely sex determination from CT-scan images of human skulls. Experiments were accomplished over an original, large-scale dataset collected on the field and involving 1400 patients. Statistical and connectionist approaches were considered. In particular, neural networks having a probabilistic interpretation of their outputs were reviewed. The two paradigms can be mixed in a variety of natural, sound ways on the basis of the probabilistic meaning of their outputs. Several combination techniques were considered and compared on the field. Results are noticeable in an application perspective, turning out to be higher than the expected correctness of prediction by human experts, as well as w.r.t. statistical approaches previously investigated in the literature on forensic sciences. In particular, combination based on committee machines do particularly fit the task. Finally, P-ANNs proved themselves to be more effective than traditional statistical techniques over the multivariate density estimation task at hand.

References

1. Bishop, C.M.: Neural Networks for Pattern Recognition. Oxford University Press, Oxford (1995)
2. Brasili, P., Toselli, S., Facchini, F.: Methodological aspects of the diagnosis of sex based on cranial metric traits. Homo. 51, 68–80 (2000)
3. Duda, R.O., Hart, P.E.: Pattern Classification and Scene Analysis. Wiley, New York (1973)
4. Haykin, S.: Neural Networks. A Comprehensive Foundation. Macmillan, New York (1994)
5. Hsiao, T.H., Chang, H.P., Liu, K.M.: Sex determination by discriminant function analysis of lateral radiographic cephalometry. Journal of Forensic Sciences 41(5), 792 (1996)
6. Nixon, M., Aguado, A.S.: Feature Extraction & Image Processing, 2nd edn. Academic Press (2008)
7. Novotny, V., Iscan, M., Loth, S.: Morphologic and osteometric assessment of age, sex, and race from the skull. In: Iscan, M.Y., Helmer, R.P. (eds.) Forensic Analysis of the Skull, pp. 71–88. Wiley-Liss, New York (1993)
8. Rsing, F.W., Graw, M., Marr, B., Ritz-Timme, S., Rothschild, M.A., Rzscher, K., Schmeling, A., Schrder, I., Geserick, G.: Recommendations for the forensic diagnosis of sex and age from skeletons. HOMO - Journal of Comparative Human Biology 58(1), 75–89 (2007)
9. Trentin, E.: Networks with trainable amplitude of activation functions. Neural Networks 14(4-5), 471–493 (2001)
10. Trentin, E.: Simple and Effective Connectionist Nonparametric Estimation of Probability Density Functions. In: Schwenker, F., Marinai, S. (eds.) ANNPR 2006. LNCS (LNAI), vol. 4087, pp. 1–10. Springer, Heidelberg (2006)
11. Walrath, D.E., Turner, P., Bruzek, J.: Reliability test of the visual assessment of cranial traits for sex determination. American Journal of Physical Anthropology 125(2), 132–137 (2004)
12. Zhang, D., Lu, G.: A Comparative Study on Shape Retrieval Using Fourier Descriptors with Different Shape Signatures. Journal of Visual Communication and Image Representation 14(1), 41–60 (2003)

Classification of Emotional States in a Woz Scenario Exploiting Labeled and Unlabeled Bio-physiological Data

Martin Schels[1], Markus Kächele[1], David Hrabal[2], Steffen Walter[2], Harald C. Traue[2], and Friedhelm Schwenker[1]

[1] Institute of Neural Information Processing, University of Ulm, Germany
[2] Medical Psychology, University of Ulm, Germany
firstname.lastname@uni-ulm.de

Abstract. In this paper, a partially supervised machine learning approach is proposed for the recognition of emotional user states in HCI from bio-physiological data. To do so, an unsupervised learning preprocessing step is integrated into the training of a classifier. This makes it feasible to utilize unlabeled data or – as it is conducted in this study – data that is labeled in others than the considered categories. Thus, the data is transformed into a new representation and a standard classifier approach is subsequently applied. Experimental evidences that such an approach is beneficial in this particular setting is provided using classification experiments. Finally, the results are discussed and arguments when such an partially supervised approach is promising to yield robust and increased classification performances are given.

1   Introduction and Related Work

The reliability of a classifier heavily depends on the quality and quantity of the data, that was available for its construction. Unfortunately, in real world applications, it is often not trivial to design data bases where the data samples are exhaustively labeled. The main reason for this is that the general procedure of labeling data is often time consuming and expensive as it requires the knowledge of human experts.

There are several techniques in the literature, that aim at circumventing this issue by incorporating a machine-conducted labeling procedure: to make the annotation process more effectively, active learning is often used to guide a human expert during an annotation process. Hereby, the most informative sample from the unlabeled data, i.e. the one closest to a precomputed decision boundary, is selected by the algorithm and passed to an expert [4]. In order to conduct a fully automatic process, semi supervised learning can be applied: classifiers are directly used to annotate the unlabeled data. A classifier can label data for itself by choosing the most confident data samples and add them to the training set (self training) [15]. Another option is to use several classifiers in order to mutually select confident samples for the respective training data (co-training) [2,6,9].

F. Schwenker and E. Trentin (Eds.): PSL 2011, LNAI 7081, pp. 138–147, 2012.

In this contribution, we implement a further learning strategy to exploit unlabeled data in a classification process. The key idea is to infer the general structure of the application using methods of unsupervised learning [25]. This leads to a representation space using the cluster centers as reference system. For these computations all available data can be used. Such an approach appeared to be beneficial in previous work such as [19].

The remainder of this paper is organized as follows: The underlying data collection is described in Section 2 together with the employed features. Section 3 points to the general issues that occur in the application and introduces the proposed method in greater detail. The experiments and the respective results are shown in Section 4. Finally, in Section 5 these results are discussed and conclusions are drawn.

2   Data Collection

The data was collected in a Wizard-of-Oz study [7], which was conducted in order to investigate affective human computer interaction in the well established PAD space. The PAD model [17] defines a three dimensional annotation scheme of emotions using the three dimensions *pleasure*, *arousal* and *dominance*.

In this particular setting, the test persons were instructed to solve multiple games of concentration using a voice controlled interface. The successive games were used to induce different emotional states to the subject in the order sketched in Figure 1. To do so, different stimuli were presented to the subject deliberately: Different negative (dispraise, time pressure, wrongly or delayed executed commands, etc.) as well as positive (e.g. praise, easier game) behaviors of the computer interlocutor were presented. The subjects were passed through 5 sequences, which induce different states in the PAD space and the subjects each passed through these sequences twice in two successive sessions (see Figure 1 for details) [24]. Each of these sequences has a length of 3-5 minutes. Overall, 20 subjects (21 to 55 years, 10 male and 10 female) were passed through the experimental procedure twice and thus for every person two experimental sequences are available.

As a whole, 5 different channels were recorded at a sample rate of 512 Hz, namely blood volume pulse (i.e. heart rate), electromyography (attached to musculus zygomaticus and musculus currugator), skin-conductance and respiration. From these signals, various features were extracted on different time scales. Hereby, it is crucial to conduct a careful preprocessing procedure in order to remove artifacts but to retain the respective information. In general, a slow low- or band-pass filter is applied together with a linear piece-wise detrend[1] of the time series at a 10 s basis. In the following, a list of the extracted features per channel is provided. The preprocessing together with the time granularity is given in parentheses.

---

[1] i.e. subtracting piecewise a linear least-squares-fit from the respective chunk of the data.

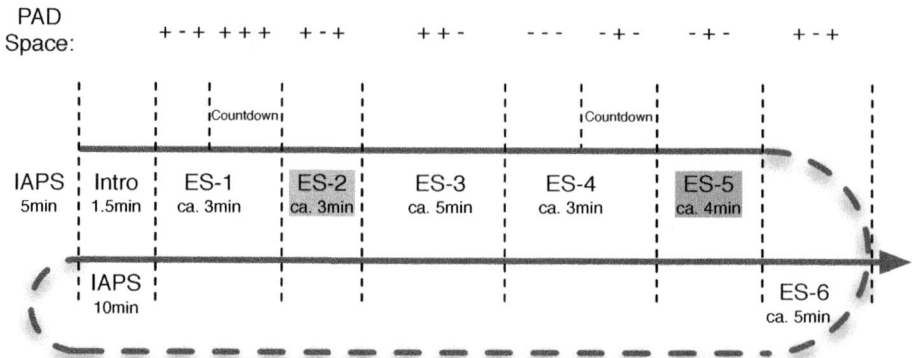

Fig. 1. Experimental design, including the expected position in the space. Experimental sequences ES-2 and ES-5, marked green and red respectively, are expected to induce the desired emotional states [24]. The top row in the figure indicates the intended label in PAD Space, whereby "+" signifies a high value in the respective dimension and "-" vice versa.

Blood volume pulse (BVP) is recorded from an optical sensor device, attached to a finger of the subject. The key to characterize the heart rate from the recorded blood volume pulse is to find the well known QRS complex in the signal e.g. as described in [16]. The following features are extracted (low pass filtered at 5 Hz, 25 s time window each) : Standard deviation of heart rate variability [18], standard deviation of RR-intervals [23], pNN50[2] [12], approximate entropy [13], RMSSD[3] and recurrence rate, Poincaré plot[4] [10] and power spectral density [26] of the signal.

The subject's respiration [3] is measured using a belt, that is wrapped around its breast and has a tension measurement device attached to. From this signal the following features are computed (low pass filtered at 0.15 Hz): Mean and standard deviation of the first derivatives (10 s time window), breathing volume, mean and standard deviation of breath intervals, Poincaré plot[4] (30 s time window each).

To record the electromyogram (EMG), 2 electrodes are attached to the skin near to the respective muscle. Thus electrical potential differences of about 500 μV are recorded. Hereby lies the information of contraction or relaxation of the muscle in the oscillation of the EMG signal. The following features were computed (bandpass filtered at 20 - 120 Hz, piecewise linear detrend): Mean of first and second derivatives (5 s time window), power spectrum density estimation [26] (15 s time window).

---

[2] The pNN50-measure equals the proportion of occurrences of changes in RR-interval duration of two consecutive RR-intervals that differ more than 50ms.

[3] Square root of the mean of squared successive differences of RR-intervals.

[4] Ratio of the axes of the fitted ellipse.

Skin conductance is measured (SCL) using 2 electrodes, where constant electrical current of 10 μA conducted. The respective resistance is then determined by the sweat, a subject oozes. The following features are extracted for this signal (low pass filtered at 0.2 Hz): mean and standard deviation of first and second derivative (5 s time window), mean peak occurrences, average peak-height (20 s time window each) [5].

In the following, the task to solve in the context of this paper is to discriminate the samples of ES-2 and ES-5. These two experimental sequences are designed to elicit rather complementary emotions: "high pleasure/low arousal/high dominance" versus "low pleasure/high arousal/low dominance" (compare Figure 1, top row) – or short positive vs. negative emotion. The according stimuli that are presented to the subject were praises and a small board of concentration and hence an easy play for the positive sequence. In case of the negative class, the user is given a bigger board and only displeasing feedback is given: e.g. the user is criticized for his execution of the game and the subject is exposed to time pressure.

3   Problem Statement and Proposed Method

An application as described above arises several severe issues from a machine learning perspective. Based on the design of the psychological experiment, the overall samples that are labeled accordingly are very rare. When attempting to compute reasonable features from the given data, the respective time window has to be chosen over several seconds. Due to the high differences of physiology over different subjects, the given application encourages the commands for a personalized setting in the training of classifiers. This further toughens the lack of data.

Further, when evaluating such kind of data it is not recommended to use some kind of "leave one sample out" technique to evaluate a statistical model. The employed sensors show a distinct characteristic over time and as the labels are heavily correlated by definition to time, this would imply a severe bias in the results. This implies in our application that it is necessary to train and test using data from different sessions. Hence, it is highly desirable to make use of all available data from all experimental sequences recorded from a subject. Unfortunately this data is not labeled in the respective classes (compare Figure 1). On the other hand, it is still data from the same domain. The goal is now to incorporate all available data into the construction of a classifier for the considered two classes.

To do so, it is rather intuitive to refer to techniques of unsupervised learning. The key idea is now to neglect the actual class labels for the samples and to process all available data using a unsupervised technique - such as k-means or Gaussian mixture models. In order to solve the actual classification problem a further learning step is implemented: Based on the computed partitioning of the data, an "activation value" of the cluster centers for the data samples is computed. This activation could either be computed by a distance measure with

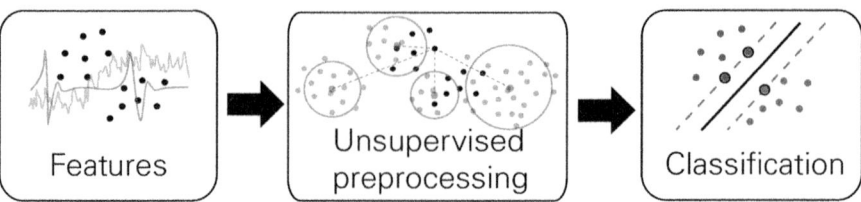

Fig. 2. In order to incorporate the unlabeled data into the classification, an unsupervised learning step is implemented. Thus, the data is transformed into a new representation, in which the actual classification is conducted.

respect to a cluster center in case of a partitioning algorithm is used, or the posterior probability of a mixture component of a fitted generative model. This results in a new representation of the data of the same dimensionality as number of cluster centers. Based on this new feature vector, a classification on the initial label is conducted using standard supervised machine learning approaches. This procedure is sketched in Algorithm 1.

Algorithm 1: Proposed algorithm in pseudo code.

Input:
- Labeled data $L = (l_i)_{i=1...M}$
- Respective labels $Y = (y_i)_{i=1...M}$
- Unlabeled data $U = (u_j)_{j=1...O}$
- Number of cluster centers k,

compute k local densities or prototypes $p_1, \ldots, p_k$ using $L \cup U$;
foreach $l_i \in L$ **do**

$\quad l'_i = G_{p_1,\ldots,p_k}(l_i) \in \mathbb{R}^N$;

$\quad G$ is a distance or similarity measure, and N is a natural number depending on the specific structure of G

$\quad$ examples:

$\quad$ (a) $N = k$, and $G_{p_1,\ldots,p_k}(l) = | p_i - l |$
$\quad$ (b) $N = k$, and $G_{p_1,\ldots,p_k}(l) = (\exp(- | p_i - l | /\sigma_i)$
$\quad$ (c) $N = k(k-1)/2$ and $G_{p_1,\ldots,p_k}(l_i) = \min_{i,j}(| p_i - l |, | p_j - l |)$

end
Train classifier F on $((l'_i)_{i=1...M}, Y)$;
Output: F

To classify an unseen data sample, it has to be transformed into the new representation. This is done analogously to the training procedure by calculating the activation score: These values are computed with respect to the computed local density or the respective prototype and the obtained new representation is classified.

4   Experiments and Results

In this section the conducted experiments and the obtained results are described. An important step in our approach is the choice of the unsupervised learner: We decided to evaluate the well known k-means algorithm, neural gas [11] and Gaussian mixture models (GMM) trained through the well known expectation maximization (EM) algorithm. In case of neural gas and k-means, the number of cluster centers is chosen to be 15% of the size of the training set. Due to numerical issues the number of mixture components for the GMM is set constantly to 4 and a regularization constant of 0.001 was fixed. Generally, the euclidean distances to the cluster centers were used as new representation except for GMM, where the posteriori probability for every Gaussian mixture component was computed.

For the supervised part of the proposed architecture, a support vector machine (SVM) learning approach was used [1]. To be precise, in this work ν-SVM as described in [20] was used using an RBF kernel function. To compare the experimental results, we also conducted a purely supervised reference experiment, where the training of the classifier is conducted only on the accordingly labeled data. For this experiment, we used the ν-SVM approach with an RFB kernel as well.

In Section 3 the general issue of testing a classifier in this application appropriately was mentioned briefly. To circumvent this issue and in order to ensure proper results, the partitioning in training and testing data are separated by session per class. But to increase the possible settings for testing, the whole experimental sequences are permuted in all possible combinations. This leads to four settings of testing and training sets per subject.

The features described in Section 2 are extracted not only from different modalities but also in different time scales. Hence, the classification study was conducted in six different experiments, grouping the data by feature and size of the time window: For the EMG, features that govern in time domain (derivatives of the signal and related) are grouped together as well as features obtained from the power spectrum. Also for the skin conductance two groups of features were defined for classification: The statistics over the derivations are processed in a different classifier than the statistics of the peaks of the signals. In case of BVP and respiration such a partitioning is not necessary as the time windows of all extracted features are the same.

The performances of the classifiers are reported in Table 1. As the distribution of classes in the data is imbalanced (compare 1) not only the accuracy are reported, but also the F1 scores for ES-2. The numbers in the rows are mean values over all subjects and every classifier is evaluated 80 times each. Generally, the numbers are relatively low, which is not surprising as the application is rather challenging together with the general lack of data. It can be observed that the two classifiers using EMG features perform best with an accuracy up to 0.53. Also the classification on respiration features performs well (0.51 accuracy). All classifiers avoid to produce one-sided classification result, i.e. it does not constantly decide for the class having the higher a priori probability, which is indicated by the F1 scores.

Table 1. Accuracies and F1 scores for ES-2 for the 6 classifier configurations averaged over all subjects and 80 trials per subjects. The row-wise maximal values for both values are highlighted in bold font. The last row shows the averages over all classification trials.

Feature combination	GMM		Neural-Gas		K-Means		purely supervised	
	acc.	F1	acc.	F1	acc.	F1	acc.	F1
EMG (derivatives)	0.529	0.404	**0.530**	**0.428**	0.486	0.389	0.502	0.394
EMG (power spectrum)	**0.531**	0.404	0.514	**0.430**	0.461	0.366	0.458	0.342
SCL (derivatives)	0.431	**0.325**	**0.431**	0.300	0.415	0.321	0.424	0.323
SCL (inter-peak statistics)	0.437	0.356	0.448	0.399	**0.451**	**0.399**	0.421	0.355
BVP	0.475	0.363	0.437	0.355	0.455	0.366	**0.483**	**0.392**
Respiration	0.449	0.347	**0.510**	**0.447**	0.484	0.407	0.503	0.368
Average	0.475	0.366	**0.479**	**0.393**	0.459	0.374	0.465	0.362

Table 1 provides some arguments, that the unsupervised preprocessing does provide benefits for the classification: In 3 of 6 cases of the classifiers, the partly supervised method using neural gas outperforms the others. Further, comparing all partly supervised experiments to the purely supervised case, it performs best in 5 of 6 cases on average. Also, when averaging over all test runs, there is a slight preference for the clustering preprocessing approach using neural gas but also using GMM.

A ranking-like experiment is conducted, where it is counted how often a classifier outperforms all others for every individual subject averaged over all 80 trials. The results of this are reported in Table 2 as fractions of all comparisons. This consideration reveals a slight advantage of the GMM based partially supervised classifier: It outperforms the others in 32% of the cases. Especially for the features from EMG, which performed best in Table 1, such an approach appears to be beneficial.

Table 2. For every classifier it is shown how often it outperforms all others. The lines of the table show different feature combinations.

Feature combination	GMM	Neural-Gas	K-Means	purely supervised
EMG (derivatives)	**41.1%**	17.7%	17.7%	23.5%
EMG (power spectrum)	**47.1%**	23.5%	5.9%	23.5%
SCL (derivatives)	23.5%	17.7%	23.5%	**35.3%**
SCL (inter-peak statistics)	29.4%	17.7%	**35.3%**	17.7%
BVP	**35.3%**	5.9%	23.5%	**35.3%**
Respiration	17.7%	**35.3%**	11.8%	**35.3%**
Average	**32.4%**	19.6%	19.6%	28.4%

5   Discussion and Future Work

In this work a partially supervised machine learning approach has been proposed and applied to the classification of bio-physiological time series. In this application, only few data is available in the considered classes, but there is differently annotated data at hand, that did arise in the overall recording process. The goal was to incorporate these samples into the classification process. To do so, we propose to use an unsupervised learning approach as a preprocessing step. Three different learning strategies have been evaluated in this context: k-means, neural gas as clustering approaches and GMM to estimate the probability distribution. Thus the data was transformed in a new representation using the activation per prototype or mixture component. Using the partitioning algorithm, the euclidean distance has been used, while for the GMM the posterior probability per mixture component is used. The experimental results reveal a slight advantage over the purely supervised reference method of such an approach in this application.

In order to provide a rationale of why the proposed method works, the reader is pointed to the the well known RBF networks. There exists a big research community exploring how to improve the training of a network from given data by finding a proper initialization [8,14,21,22]. Hereby, the aim is to make the results more stable and also to speed up training. A typical approach is to pretrain the hidden RBF-layer in an unsupervised fashion by clustering or vector quantization. Afterwards, the network is finally trained by either solely creating a perceptron for the output layer or back propagation for the whole network. The unsupervised step in our approach can be regarded as some sort of initialization of a "hidden layer" using all available data. Thus, the distributions of data can be estimated more reliably. After that, a second "output layer" is created with only the labeled data at hand.

Adding additional data the way we did in our experiments, i.e. data, that is not from the same categories is of course only promising under certain conditions. If the samples of data resolved into clearly delimited classes, where the probability density functions for the different categories are non-overlapping, adding data from a very different partition would hardly be reasonable. But in many real world applications, this optimal setting for a classifier is not really present: Often the data decomposes into severely overlapping distributions. There are also applications, where the particular classes are not (yet) irrevocably defined or such a definition is simply not possible due to distinct properties. Both circumstances are at hand in the application described earlier: On the one hand, the features that can be extracted from the bio-signals can be considered relatively weak compared to the intended – quite ambitious – objective. On the other hand, even though the induction of the intended emotion succeeds in the average, it is not guaranteed by any means that every particular sample is correctly labeled.

The relatively small accuracies reported in Table 1 could be regarded as a major flaw of this contribution. There are two major ways to heal this issue: There are still 6 individual classifiers that are evaluated in this study. These classifiers should be further combined in order to enhance a frame wise classification process. This has to include of course some kind of alignment of the different time windows that are used in order to get a coherent classification. Another promising approach is to integrate the decisions of the classifiers over time [24]. On the other hand, one will then have to solve a general segmentation problem in order to discriminate the sequences.

Further, calculating the new representation of the data samples creates the opportunity to define mappings into higher dimensional spaces. This could, for example, conducted as sketched in Algorithm 1 at example (c), where pair-wise comparisons are used to build the new representation. Thus, it might be more likely to find a proper linear separation of the respective classes.

Acknowledgments. This paper is based on work done within the Transregional Collaborative Research Centre SFB/TRR 62 "Companion-Technology for Cognitive Technical Systems", funded by the German Research Foundation (DFG). The work of Martin Schels is supported by a scholarship of the Carl-Zeiss Foundation.

References

1. Bennett, K.P., Campbell, C.: Support vector machines: hype or hallelujah? SIGKDD Explor. Newsl. 2, 1–13 (2000)
2. Blum, A., Mitchell, T.: Combining labeled and unlabeled data with co-training. In: Proceedings of the Eleventh Annual Conference on Computational Learning Theory, pp. 92–100 (1998)
3. Boiten, F.A., Frijda, N.H., Wientjes, C.J.: Emotions and respiratory patterns: review and critical analysis. International Journal of Psychophysiology 17(2), 103–128 (1994)
4. Cohn, D.A., Ghahramani, Z., Jordan, M.I.: Active learning with statistical models. J. Artif. Int. Res. 4, 129–145 (1996)
5. Darrow, C.W.: The equation of the galvanic skin reflex curve: I. the dynamics of reaction in relation to excitation-background. The Journal of General Psychology 16(2), 285–309 (1937)
6. Hady, M.F.A., Schwenker, F., Palm, G.: Semi-supervised learning for tree-structured ensembles of RBF networks with co-training. Neural Networks 23(4), 497–509 (2010)
7. Kelley, J.F.: An empirical methodology for writing user-friendly natural language computer applications. In: Proceedings of the SIGCHI conference on Human Factors in Computing Systems, pp. 193–196 (1983)
8. Kestler, H.A., Schwenker, F., Hoher, M., Palm, G.: Adaptive class-specific partitioning as a means of initializing RBF-networks. In: IEEE International Conference on Systems, Man and Cybernetics, vol. 1, pp. 46–49 (1995)
9. Ling, C.X., Du, J., Zhou, Z.H.: When does co-training work in real data? In: Proceedings of the 13th Pacific-Asia Conference on Advances in Knowledge Discovery and Data Mining, pp. 596–603 (2009)
10. Marciano, F., Migaux, M., Acanfora, D., Furgi, G., Rengo, F.: Quantification of Poincaré maps for the evaluation of heart rate variability. Computers in Cardiology, 577–580 (1994)
11. Martinetz, T., Schulten, K.: A "Neural-Gas" Network Learns Topologies. Artificial Neural Networks I, 397–402 (1991)
12. Mietus, J.E., Peng, C.K., Goldsmith, R.L., Goldberger, A.L.: The pNNx files: re-examining a widely used heart rate variability measure. Heart 8, 378–380 (2007)

13. Pincus, S.M.: Approximate entropy as a measure of system complexity. Proceedings of the National Academy of Sciences of the United States of America 88(6), 2297–2301 (1991)
14. Ros, F., Pintore, M., Deman, A., Chrtien, J.R.: Automatical initialization of RBF neural networks. Chemometrics and Intelligent Laboratory Systems 87(1), 26–32 (2007)
15. Rosenberg, C., Hebert, M., Schneiderman, H.: Semi-supervised self-training of object detection models. In: Proceedings of the Seventh IEEE Workshops on Application of Computer Vision, pp. 29–36 (2005)
16. Rudnicki, M., Strumiłło, P.: A Real-Time Adaptive Wavelet Transform-Based QRS Complex Detector. In: Beliczynski, B., Dzielinski, A., Iwanowski, M., Ribeiro, B. (eds.) ICANNGA 2007. LNCS, vol. 4432, pp. 281–289. Springer, Heidelberg (2007)
17. Russell, J.: A circumplex model of affect. Journal of Personality and Social Psychology 39, 1161–1178 (1980)
18. Sayers, B.: Analysis of heart rate variability. Ergonomics 16(1), 17–32 (1973)
19. Schels, M., Schillinger, P., Schwenker, F.: Training of multiple classifier systems utilizing partially labeled sequences. In: Proceedings of the 19th European Symposium on Artificial Neural Networks, pp. 71–76 (2011)
20. Schölkopf, B., Smola, A.J., Williamson, R.C., Bartlett, P.L.: New support vector algorithms. Neural Comput. 12, 1207–1245 (2000)
21. Schwenker, F., Kestler, H., Palm, G., Hoher, M.: Similarities of LVQ and RBF learning-a survey of learning rules and the application to the classification of signals from high-resolution electrocardiography. In: IEEE International Conference on Systems, Man, and Cybernetics, vol. 1, pp. 646–651 (1994)
22. Schwenker, F., Kestler, H.A., Palm, G.: Three learning phases for radial-basis-function networks. Neural Netw. 14, 439–458 (2001)
23. Simson, M.: Use of signals in the terminal QRS complex to identify patients with ventricular tachycardia after myocardial infarction. Circulation 64(2), 235–242 (1981)
24. Walter, S., Scherer, S., Schels, M., Glodek, M., Hrabal, D., Schmidt, M., Böck, R., Limbrecht, K., Traue, H.C., Schwenker, F.: Multimodal Emotion Classification in Naturalistic user Behavior. In: Jacko, J.A. (ed.) HCI International 2011, Part III. LNCS, vol. 6763, pp. 603–611. Springer, Heidelberg (2011)
25. Webb, A.R.: Statistical Pattern Recognition. John Wiley and Sons Ltd. (2002)
26. Welch, P.: The use of fast Fourier transform for the estimation of power spectra: A method based on time averaging over short, modified periodograms. IEEE Transactions on Audio and Electroacoustics 15(2), 70–73 (1967)

Using Self Organizing Maps
to Find Good Comparison Universities

Cameron Cooper[1] and Robert Kilmer[2]

[1] Fort Lewis College, Freshman Mathematics Program, Durango, Colorado, United States
cooper_c@fortlewis.edu
[2] Walden University, School of Management, Minneapolis, Minnesota, United States
robert.kilmer@waldenu.edu

Abstract. Colleges and universities do not operate in a vacuum and they do not have a lock on "best practices". As a result it is important to have other schools to use for "benchmark" comparisons. At the same time schools and their students change. What might have been good "benchmarks" in the past might not be appropriate in the future. This research demonstrates the viability of Self Organizing Maps (SOMs) as a means to find comparable institutions across many variables. An example of the approach shows which schools in the Council of Public Liberal Arts Colleges might be the best "benchmarks" for Fort Lewis College.

Keywords: Kohonen self organizing maps, neural networks, benchmarking, higher education.

1 Introduction

Competitive organizations, such as institutions of higher education, continuously need to compare how well they do in relation to other organizations of similar size, shape and function across many different dimensions. These institutions could benefit from a methodology to identify organizations that are most "like" them. The example presented in this research utilizes a publically available dataset [1], self-organized maps, an unsupervised artificial neural network, to cluster/map 25 schools across 28 variables into seven distinct clusters with an easy to interpret 2-dimensional visual map of the schools and their corresponding peers. Schools within the same clusters (usually four to five other schools) could be treated as "peer" institutions for benchmarking purposes.

2 Background

Benchmarking has been used in higher education for over two decades. [2] As a means to find good comparison universities, the authors opted to utilize a Self Organizing Map. SOMs were initially crafted by Finnish professor Tuevo Kohonen in the 1960s. A comprehensive mathematical description of Kohonen's work can be found in his book *Self-Organizing Maps* [3]. Researchers have found SOMs to be

F. Schwenker and E. Trentin (Eds.): PSL 2011, LNAI 7081, pp. 148–153, 2012.

useful in medicine for visualization of gene expression [4], [5] and numerous other medical applications. In business, they have been employed for financial benchmarking [6] and strategic positioning [7]. Additionally, SOMs create means for analysis via multidimensional scaling of highly complex data sets [8], a key feature of SOMs. The primary author also applied SOMs to identify benchmark universities on the basis of student assessment of university websites [9]. The purpose of this research is twofold. First, the SOM stemming from this research offers strategic positioning information for colleges and universities. Second, via multidimensional scaling, a key feature of SOMs, this research consolidates a 28 variable input space down to an easy to interpret two-dimensional visual map.

3   Methods

The Data Profile of the member institutions of the Council of Public Liberal arts Colleges offers a nearly complete dataset [1], one important requirement to creating a reliable SOM. Thus, the data from "Section I: Admissions and Student Characteristics" of the Data Profile [1], comprised the dataset to form a 5 X 5 Self-Organizing Map (SOM). The variables for the 25 institutions are listed Table 1.

Table 1. Admissions and Student Characteristics Variables

Variables from Section I: Admissions and Student Characteristics		
1. number of applications received	10. average ACT Composite score	20. percent out-of-state undergraduates
2. percent applicants admitted	11. average SAT Critical Reading score	21. percent undergraduates age 25 or older
3. percent admitted applicants who enrolled	12. average SAT Writing score	22. percent undergraduates living on campus
4. Number (headcount) of all full-time first-time freshmen	13. average SAT Math score	23. Number of new undergraduates who were transfer students
5. percent full-time first-time female freshmen	14. Number (headcount) of all undergraduates	24. Number (headcount) of graduate students
6. percent full-time first-time male freshmen	15. percent full-time undergraduates	25. Full-time equivalent (FTE) of all students*
7. percent minority full-time first-time freshmen	16. percent minority undergraduates	26. undergraduate FTE enrollment*
8. Number (headcount) of all part-time first-time freshmen	17. percent female undergraduates	
9. percent in top 10% of high school class	18. percent male undergraduates	27. graduate FTE enrollment
	19. percent in-state undergraduates	

The authors created a two-dimensional self organizing map via a Microsoft Excel add-in, SOMinExcel [10]. The following four steps comprised the mapping process:

1) The application initially assigns random weight vectors to the neuron centers of the 25 nodes/neurons.
2) The application calculates the Euclidean distance between all of the weight vectors and a presented observation. After the best matching unit is found (i.e. the closest distance between an observation/school and the weight vector). The school is assigned to the node with an appropriate distance from the center.
3) Then, the weight vector of the winning neuron and its neighbors are modified to bring their measures closer to the observation vector. The researchers employed a Gaussian neighborhood function for adjusting the weights of the neighboring nodes. Training comprised $\lambda=200$ cycles and ultimately formed the weighted SOM. Gaussian neighborhood updates allow the weights of the neurons closest to the winning neuron to be updated to become more similar to the weights of the winning neuron. Per the Gaussian function, the magnitude of weight modification tapers off for neurons further away from the winning neuron.
4) In the end, the software finally mapped the school records onto the SOM by determining a school's relative location within the winning neurons. This process located the plot points on the 5 X 5 map.

4   Results

Seven clusters with an average cluster size of four COPLAC member institutions resulted (see Figure 1). With the goal of finding good comparison schools, a cluster size of four offers schools' administrations a reasonable number of comparison universities.

With a two-dimensional map, institutions of higher education interested in benchmarking their operations (e.g. retention and recruitment efforts) need to only locate their school's observation number (i.e. point on the map labeled by observation number from Table 3), and find schools closest to their point in terms of Euclidean distance.

5   Discussion

For example, Fort Lewis College (Observation #2 in Figure 1), the first author's home institution, might compare itself to schools #6 (Mass. College of Liberal Arts), #4 (Henderson State University), #10 (Shepard University) and #24 (The Univ. of Virginia's College at Wise), which are within the same winning neuron. By performing a radial search from the school's labeled observation point in the SOM (see Figure 1), one would find Fort Lewis College to be most similar to Observation #4, Henderson State University. This, however, does not necessarily mean Fort Lewis College would want to model and/or compare its operations after Henderson. Per a school criteria created by administrators and school-wide initiatives, one could manually create a weighted ranking of the potential comparison schools within the same neuron.

As an example, administrators at Fort Lewis College might identify the following five variables to be the most important to improve in relation to strategic initiatives:

1. % Minority Students – percent minority full-time first-time freshmen
2. % top 10% of high school class – percent in top 10% of high school class
3. % In-state – percent in-state undergraduates
4. % 25 > age – percent undergraduates age 25 or older
5. % on campus – percent undergraduates living on campus

A weighted score out of 100 could then be created in order to rank the institutions in (Table 2) reference to desired comparability. A sample weighting with % of top 10% of high school class (i.e. 35 out of the 100 points) as the foremost priority followed by increasing the % minority (i.e. 25 points) at the institution. With these weighted priorities, it would behoove Fort Lewis College to contact Henderson State University and Massachusetts College of Liberal Arts regarding admission operations.

Per the SOM, given the innate similarities between Fort Lewis College (Observation #2) and Henderson State University (Observation #4) in regards to Admissions and Student Characteristics, one could hypothesize strategic initiatives successful at one institution would also be successful for the other.

Fort Lewis College, Henderson State University, and University and Massachusetts College of Liberal Arts are all public institutions located in different geographical areas of the United States (i.e. Colorado, Arkansas, and Massachusetts). Since they are unlikely to be competing for the same applicant pool, Henderson's Office of Admission and Massachusetts College's Office of Admission might be willing to share some of its more effective "best practices" in recruiting students from the top 10% of their graduating class.

Specifically, Fort Lewis' Office of Admission might be interested in contacting their departmental counterpart at Henderson State University in bettering its recruitment of students "in top 10% of [their] high school class", variable #9. 16% of Henderson State University's students and 21% of Massachusetts' College's students are from the top 10% graduating class whereas 7% of Fort Lewis College's students are from the top 10% of their high school class. If these benchmark schools could offer Fort Lewis College, one or two in-state recruitment strategies, then the dividends to Fort Lewis could be substantial. An increased academically prepared student body will inevitably result in higher retention rates and graduation rates, both viewed as high priority strategic initiatives supported by not only higher administration but also the board of trustees.

Table 2. Weighted Averages of Institutions within Winning Neuron

Weighting	25	35	15	15	10	Weight out of 100
Institution	%Minority	% Top 10%	% In-state	% 25 > age	% on campus	Weighted Score
Fort Lewis College	26	7	69	15	36	25.15
Henderson State Univ.	40	16	85	17	42	35.1
Mass. College of Liberal Arts	12	21	76	17	68	31.205
Shepherd Univ.	14	21	55	18	29	24.805
The Univ. of Virginia's College at Wise	18	14	95	24	35	30.75

Table 3. COPLAC Member Institutions

Obs #	School Name	Obs #	School Name	Obs #	School Name
1	The Evergreen State College	9	Ramapo College of New Jersey	18	Truman State Univ.
2	Fort Lewis College	10	Shepherd Univ.	19	Univ. of Mary Washington
3	Georgia College & State Univ.	11	Sonoma State Univ.	20	Univ. of Minn., Morris
4	Henderson State Univ.	12	Southern Oregon Univ.	21	Univ. of Montevallo
5	Keene State College	13	St. Mary's College of Maryland	22	UNC Asheville
6	Mass. College of Liberal Arts	14	SUNY College at Geneseo	23	Univ. of Science and Arts of Okla.
7	Midwestern State Univ.	15	Truman State Univ.	24	The Univ. of Virginia's College at Wise
8	New College of Florida	16	Univ. of Alberta Augustana Campus	25	Univ. of Wisconsin- Superior
		17	Univ. of IL Springfield		

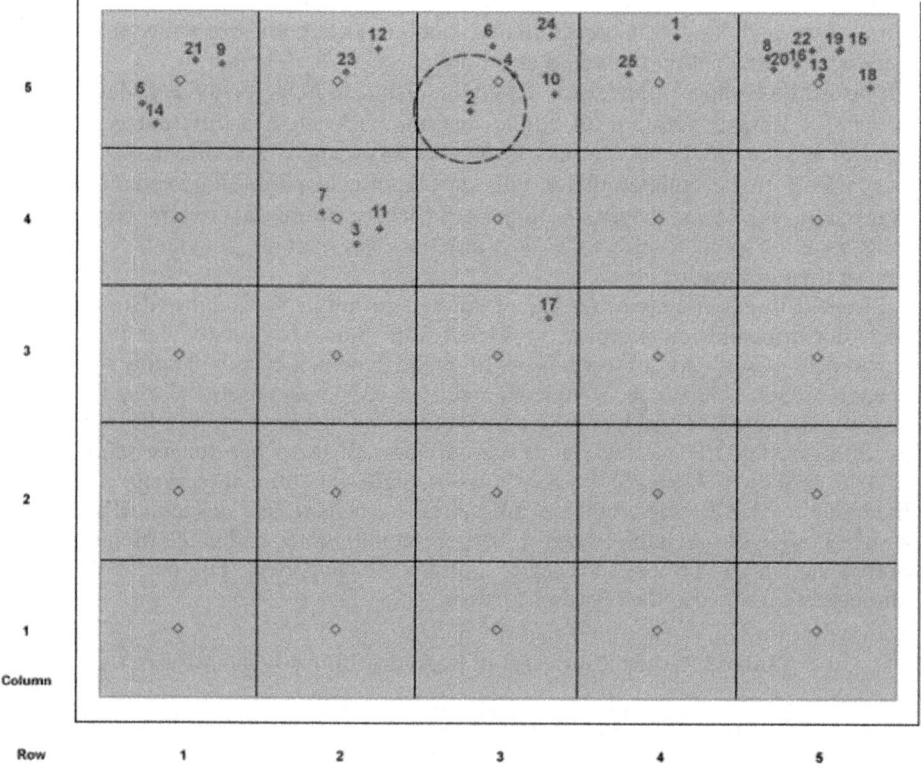

Fig. 1. Radial Search from Observation

6 Conclusions

SOMs appear to offer institutional personnel an effective, efficient, and unbiased means to discover benchmark institutions of higher education across many dimensions. Indeed the SOM in combination with a manual weighting of strategically important variables resulted in a very reasonable number of comparison schools. For future research, the results of the SOM could be used as inputs to a supervised learning approach where the examples of each cluster could be used as training and testing data.

References

[1] COPLAC, Data Profile (2009), http://www.coplac.org/resources.html (retrieved January 5, 2011)
[2] Benchmarking in Higher Education: Adapting Best Practices To Improve Quality. Eric Clearinghouse, Washington, DC
[3] Kohonen, T.: Self-Organizing Maps, 3rd edn. Springer, Berlin (2000)
[4] Venna, J., Kaski, S.: Comparison of Visualization Methods for an Atlas of Gene Expression Data Sets. Information Visualization 6(2) (2007)
[5] Kaski, S., Nikkilä, J., Oja, M., Venna, J., Törönen, P., Castrén, E.: Trustworthiness and metrics in visualizing similarity of gene expression. BMC Bioinformatics (2003)
[6] Visa, Eklund, T., Back, B., Vanharanta, H., Ari.: Using the Self-Organizing Map as a Visualization Tool in Financial Benchmarking. Information Visualization 2(3) (2003)
[7] Mazanec, J.: Positioning Analysis with Self-Organizing Maps – An Exploratory Study on Luxury Hotels. Cornell Hospitality Quarterly 6, 80–96 (1995)
[8] Castellani, B., Castellani, J., Spray, S.L.: Grounded Neural Networking: Modeling Complex Quantitative Data. Symbolic Interaction 26(4) (2003)
[9] Cooper, C., Burns, A.: Kohonen Self-organizing Feature Maps as a Means to Benchmark College and University Websites. Journal of Science Education and Technology 16(3), 203–211 (2007), doi:10.1007/sl0956-007-9053
[10] Saha, A.: Cited in Data Modeling for Metrology and Testing in Measurement science (2002); Pavese, F., Forbes, A.: p. 235 (2009)

Sink Web Pages in Web Application

Doru Anastasiu Popescu

Faculty of Mathematics and Computer Science,
University of Pitesti, Romania
dopopan@gmail.com

Abstract. In this paper, we present the notion of sink web pages in a
web application. These pages allow identifying a reduced scheme of the
web application, which can lead to simplifying the method of testing and
verifying the entire web application. We believe that this notion can be
useful in the partially supervised learning.

Keywords: Relation, Tag, HTML, Web Application.

1   Introduction

The results of this paper are related to web applications that contain web pages
consisting of HTML tags and are saved in files with .html or .htm extension. The
web pages can contain other elements as well, such as scripts or applets; however,
these will not be used next in the paper (for example: their testing and verifying implies using specific methods). Next, we will use for these pages the name of
web pages. On one hand, the number of web applications which are built using
this type of web pages is a very large one; on the other hand the web applications
can contain a large number of web pages. In this context, the matter of verifying
and testing the web application from the point of view of the content areas ([14],
[15], [10]) or from the point of view of the navigability in the application ([7], [9]).
A classification of the methods and models of testing and verifying is presented
in [9].

We believe that reduced scheme for a web application (presented in section 2)
can be used in other areas that utilize a large number of components, as it is the
case of partially supervised learning ([12], [13]).

The results presented in the following sections are related to the selection of
some web pages from the set of web pages of an application, which if tested and
verified assure the testing and verification of the entire application. The process
of detecting these web pages is realized through a relation among the web pages
of the web application (described in section 2). Different ways of defining this
relation have been presented in [1], [4], [5], [6].

The method of defining the pages that will be selected (named sink web pages)
will be introduced in section 3. Using sink web page notion, a reduced scheme of
the web application is created, which can be used in the process of testing and
verifying the web application.

F. Schwenker and E. Trentin (Eds.): PSL 2011, LNAI 7081, pp. 154–158, 2012.
© Springer-Verlag Berlin Heidelberg 2012

2   Defining a Relation between Two Web Pages

Next, we will consider a web application having the set of web pages P={p_1, p_2, ..., p_n} and a set TG of tags.

For any web page p_i from P, we write T_i the sequence of tags from p_i, which are not a member of TG (the order in which these are encountered is important).

Definition 1. Let TG be a set of tags, p_i and p_j two web pages from P. We say that $T_i=(T_{i1},T_{i2}, ..., T_{ia})$ is in $T_j=(T_{j1},T_{j2}, ..., T_{jb})$ as a sequence, if there exists an index k in the sequence T_j, with $T_{jk}=T_{i1}$, $T_{jk+1}=T_{i2}$, ..., $T_{jk+a-1}=T_{ia}$.

Definition 2. Let TG be a set of tags, p_i and p_j two web pages from P. We say that p_i is in relation R with p_j and we write p_i R p_j, if:
i) T_i is in T_j as a sequence;
ii) Any tag <Tg> from T_j which appears in T_i, as well, has a closing tag <\Tg> in T_j, then <\Tg> is also in T_i.

Example 1. Let us consider a web application with three web pages: P={p_1, p_2, p_3}. p_1 can be found in the file Pag1.html, p_2 in the file Pag2.html, and p_3 in the file Pag3.html (table 1). Considering:

Table 1. Pag1.html, Pag2.html and Pag3.html

Pag1.html	Pag2.html
<HTML> <HEAD>	<HTML> <HEAD>
<TITLE>Web page 1</TITLE>	<TITLE>Web page 2 </TITLE>
</HEAD> <BODY>	</HEAD> <BODY>
<P> Picture 1	<B> <P> Picture 1</P>
<IMG SRC="pic.jpg">	<IMG SRC="pic.jpg">
	<P> <IMG SRC="pic.jpeg">
</BODY> </HTML>	</BODY> </HTML>
Pag3.html	
<HTML> <HEAD>	</FONT> <IMG SRC="pic.jpg">
<TITLE>Web page 3</TITLE>	<FONT SIZE=4 COLOR=red>Picture 3
</HEAD> <BODY>	</FONT> <IMG SRC="pic.jpg">
<FONT COLOR=red>Picture 3	</BODY> </HTML>

TG={<P>,</P>,<B>,</B>,<HTML>,<HEAD>,<TITLE>,</TITLE>, </HEAD>, <BODY>, </BODY>, </HTML>} we obtain:
T1=(<IMG SRC="pic.jpg">); T2=(<IMG SRC="pic.jpg">, <IMG SRC="pic.jpeg">); T3=(<FONT COLOR=red>, </FONT>, <IMG SRC="pic.jpg">, <FONT SIZE=4 COLOR=red>, <IMG SRC="pic.jpg">, </FONT>).

According to definitions 1 and 2, we obtain only two pairs of pages with are in relation R:
p_1 R p_2 and p_1 R p_3.

Observations. 1. If TG=∅, then according to definition 2, two web pages p_i and p_j will be in relation R only if these pages contain exactly the same tags after removing the tags that are members of TG. In this case the relation R is an equivalence relation. A few results using this type of relation are presented in [4], [6], [7].

2. The tags from TG have to be written correctly, in order not to negatively influence the syntactic correctitude of the web pages.

3   The Concept of Sink Web Page

Using the notations in the previous section, we define the sink web pages for a web application with the set of web pages $P=\{p_1, p_2, ..., p_n\}$ and a set TG of tags, as below:

Definition. The web page p_i is called *sink web page*, if the following property is fulfilled:
it does not exist p_j in P, with i≠j and p_i R p_j.

Using the relation R, we can construct for the web application an oriented graph G=(X,U) as below:
X=\{1, 2, ..., n\} is the set of nodes. For a web page p_i, its index i is associated to it, where 1≤i≤n.
U=\{(i, j)| p_i R p_j, i≠j, 1≤i,j≤n\} is the set of edges.
Writing $d_+(i)=|\{(i,j) \mid (i,j) \in U\}|$, we obtain:

Proposition. If i, 1≤i≤n is a node in the oriented graph G, previously defined with $d_+(i)=0$, then p_i is a sink web page.

Example. For a web application with n=15 web pages and the relations between them given as in fig. 1 we obtain the following sink web pages: p_1, p_5, p_8, p_9, p_{12}, p_{15}. The following sink web pages are being obtained:
    Reduced scheme of the web pages which are part of a web application can be obtained. The scheme consists of two levels:
- the level of the sink web pages;
- the level of the web pages which are subordinated to the sink web pages through the relation R, the representation of this subordination being realized by arrows. For the previous example, the reduced scheme of the web pages is the following:

Observations - There can exist many reduced schemes, due to the fact that a web page can be in relation R to several sink web pages, the scheme showing only one relation from all of them.
- In the process of verifying and testing, there can only be used the sink web pages, because the tags from the other pages are included (in the same order and with the same properties) in these ones.
- The errors detected in the sink web pages must be repaired in these pages, as well as in the ones that are bound to these through the relation R, symbolized by arrows in the reduced diagram.

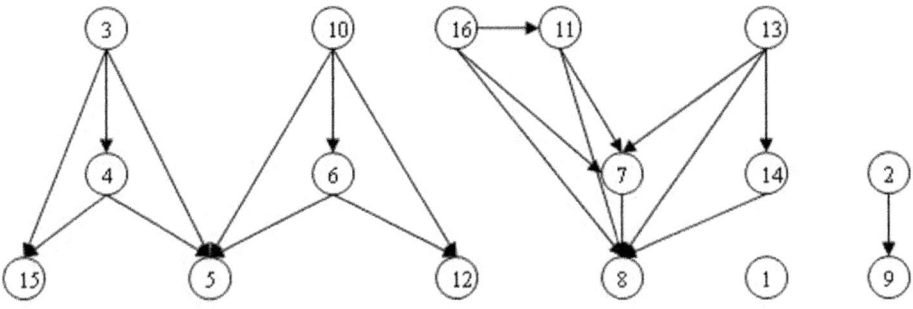

Fig. 1. The oriented graph G for a web application with 15 web pages and the relation R given by the edges

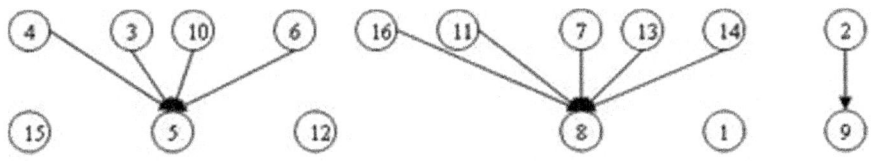

Fig. 2. A reduced scheme for the web application with 15 web pages and the relation R given in fig.1

4   Conclusion and Future Work

The concept of sink web page can be used in both processes of testing and verifying, as it was mentioned in the previous sections, but also in measuring the static complexity of the navigability ([11]) and of the content ([8], [10]) of a web application taking into consideration only these web pages. Another application of web pages sink is related to copyright, that is to simplify the comparison of two web applications, [3]. We intend to realize in the future a statistical study on categories of web applications regarding the number of sink web pages and their impact on testing, verifying, measuring, when using a complex application. In the same time we want to expand the area of applicability of the notions presented in the previous sections. The general paper from the area of partially supervised learning, like [12] and [13] show us that there is potential in this direction.

References

1. Danauta, C.M., Popescu, D.A.: Method of reduction of the web pages to be verified when validating a web site. Buletin Stiintific, Universitatea din Pitesti, Seria Matematica si Informatica (15), 19–24 (2009)
2. Cormen, T.H., Leiserson, C.E., Rivest, R.L., Stein, C.: Introduction to algorithms, 2nd edn. MIT Press (1990)

3. Popescu, D.A., Danauta, C.M.: Similarity measurement of web sites using sink web pages. In: 35th International Conference on Telecommunications and Signal Processing, pp. 24–24. IEEE Xplore (2011)
4. Popescu, D.A., Danauta, C.M., Szabo, Z.: A method of measuring the complexity of a web application from the point of view of cloning. In: Proceedings of the 5th International Conference on Virtual Learning, Section Models and Methodologies, October 29-October 31, pp. 186–181 (2010)
5. Popescu, D.A., Szabo, Z.: Sink web pages of web application. In: Proceedings of the 5th International Conference on Virtual Learning, Section Software Solutions, October 29-October 31, pp. 375–380 (2010)
6. Popescu, D.A.: A relation between web pages. In: CKS 2011: Proceedings of Challenges of the Knowledge Society, April 15-16, pp. 2026–2033. "Nicolae Titulescu" University and "Complutense" University, Bucharest (2011)
7. Popescu, D.A.: Reducing the navigation graph associated to a web application. Buletin Stiintifc - Universitatea din Pitesti, Seria Matematica si Informatica (16), 125–130 (2010)
8. Anastasiu, D.P.: Classification of links and components in web applications. In: The 2nd International Conference on Operational Research ICOR 2010, Constanta, Romania, September 09-12 (2010)
9. Alalfi, M.H., Cordy, J.R., Dean, T.R.: Modeling methods for web application verification and testing: State of art. JohnWiley and Sons, Ltd. (2008)
10. Mao, C.-Y., Lu, Y.-S.: A Method for Measuring the Structure Complexity of Web Application. Wuhan University Journal of Natural Sciences 11(1) (2006)
11. Sreedhar, G., Chari, A.A., Ramana, V.V.: Measuring Qualitz of Web Site Navigation. Journal of Theoretical and Applied Information Technology 14(2) (2010)
12. Zhu, X., Goldberg, A.B.: Introduction to Semi-Supervised Learning, Synthesis Lectures on Artificial Intelligence and Machine Learning. Morgan and Claypool (2009)
13. Zhu, X.: Semi-Supervised Learning Literature Survey. University of Wisconsin Madison (2008)
14. Alpine HTML Doctor, http://www.alpineinternet.com/
15. Validome HTML/XHTML/..., http://www.validome.org/
16. W3C Markup Validation Service, http://validator.w3.org

Author Index